SEA SERPENT CARCASSES
SCOTLAND: from the Stronsa Monster to Loch Ness

GLEN VAUDREY

978-1-905723-93-5

Typeset by Jonathan Downes,
Cover and Layout by SPiderKaT for CFZ Communications
Using Microsoft Word 2000, Microsoft Publisher 2000, Adobe Photoshop CS.

First published in Great Britain by CFZ Press

**CFZ Press
Myrtle Cottage
Woolsery
Bideford
North Devon
EX39 5QR**

© CFZ MMXII

All rights reserved. Without limiting the rights under copyright reserved above, no part of this publication may be reproduced, stored in or introduced into a retrieval system, or transmitted, in any form of by any means (electronic, mechanical, photocopying, recording or otherwise), without the prior written permission of both the copyright owners and the publishers of this book.

ISBN: 978-1-905723-93-5

Dedicated to my beautiful wife Kerry

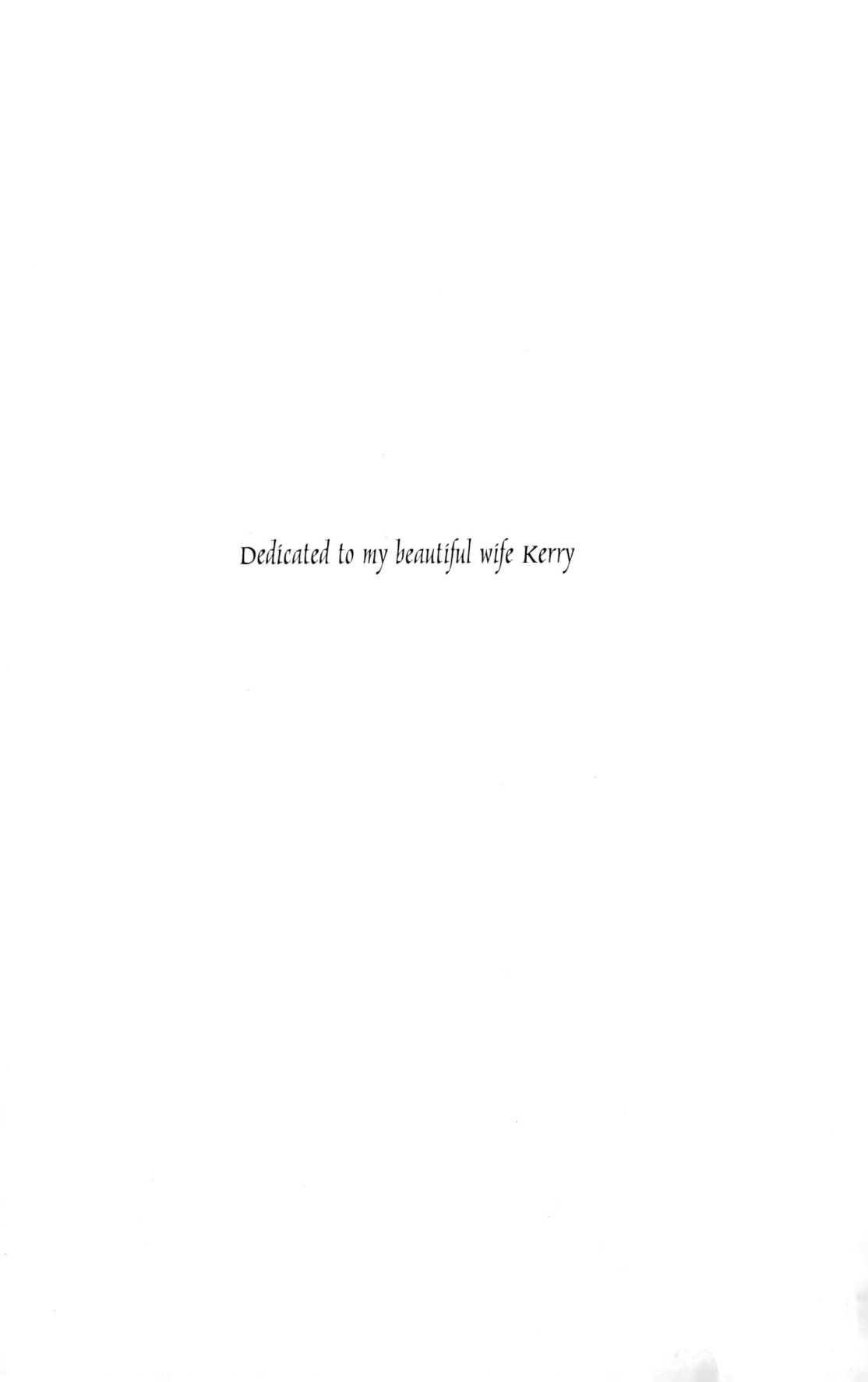

Other books by the author
Mystery Animals of the British Isles: The Western Isles
Mystery Animals of the British Isles: The Northern Isles

CONTENTS

7. Acknowledgements
8. Map of Scotland
9. Introduction
11. Alba 906
12. Stronsa Monster 1808
35. Scalloway 1810
38. Stornoway 1821
39. Benbecula 1830
43. Firth of Forth 1848
44. Usan 1849
46. Griais 1851
47. Loch Ness 1868
48. Isle of Man 1872
49. Oban 1877
53. Orkney 1894
54. Caledonian Canal 1899
55. North Atlantic Ocean 1908
57. Dunnet Sands 1934
59. Moray Firth 1934
61. Prestwick 1939-45
63. Deepdale 1941
71. Hunda 1942
75. Gourock 1942
78. Machrihanish 1944
79. Troon 1948
82. Irvine 1950s

82.	Girvan 1953
89.	Barra 1961
91.	North Sea 1963
92.	Loch Ness 1972
97.	Luce Bay 1981
99.	Benbecula 1990
102.	Loch Ness 2001-2005
104.	Bridge of Don 2011
107	Conclusions and Random Sea Serpent Musings
113	Chronology of Sightings and Related Events
117	Bibliography
119	Index

ACKNOWLEDGEMENTS

I would like to thank the following for their help and assistance:

Dr Yvonne Beale for her impressive work on the Stronsa Monster; Markus Hemmler for the fine research and leads he provided; Andreas Trottmann finder of the Luce Bay report; Karl Shuker for providing answers to many a daft question; Dale Drinnon for the tale of the Prestwick beast; Simon Welfare for keeping the mysterious world alive, and also to my family for their encouragement, and the good folks at the CFZ for letting me write this book.

INTRODUCTION

It is true to say Scotland has wonderful beaches with miles of golden sand; often you will have the beach all to yourself, but before you think this is a travel book I should add that sometimes you're sharing the beach not with a bus load of sunbathers but with a stranded sea serpent. Now you might think the chances of that happening are next to impossible, but as this book aims to prove it's not as rare as you might think.

Of course it's best to come clean at the beginning; not all of these supposed sea serpents are such. There are many large dead creatures that have washed ashore and also occasionally parts of large dead things too. There is of course a name for the latter; the term for one of these large unidentified lumps is a globster, and Scotland has had a few of these over the years.

In this book we will be looking not just at those creatures that have come ashore from the sea but also at the ones that have been found floating dead in the water around the coast, as well as having a look at the supposed remains of lake monsters. You may be surprised to learn just how many times a bit of Nessie has been found.

Where better to start but at the very beginning, back in what is known today as the Dark Ages. At that time Scotland was a very different place than it is today, for a start it was different in shape with boundaries that would not be easily recognised nowadays, while the population would be made of some very different groups, Scots, Picts, Angles and even Vikings. But one thing from back then is still to be seen occasionally today, and that is the mysterious carcass washed up on a lonely beach. As you read the book you will find that this is one thing that has carried on throughout recorded history.

Would you pay good money to look at a log like this? Read on to find out who did.

Alba 906

The very first Mystery Sea Serpent Carcass (MSSC) recorded is said to have washed ashore around the year 906.

The recorded sighting comes from the Irish Annals of Inisfallen, a chronicle of the early history of Ireland that is believed to have been written some time between the 12th and 15th Centuries by the monks of Innisfallen Abbey. It largely charts events in Ireland between AD433 and AD1450, but sometimes it covered things of interest in different countries and that is exactly what it does in a rather cryptic entry for 906.

> A woman was cast up on the shore of Alba this year. Her length [was] 192 feet; the length of her plaits 16 [feet]; the length of the fingers of her hand 6 feet; the length of her nose 6 [feet]; her body as white as a swan or the foam of a wave.

Well what are we to make of that then?

The first thing to consider is whether a very large lady could have washed ashore. While worldwide mythology does suggest that a race of giants could be found wandering the earth at some time in the past, there isn't all that much evidence of such folk wandering around Scotland. Certainly within Scottish legend we have a few reported giants, the Big Man of Ben Macdui is stated to reach around 10 foot in height, which would make him a few feet shorter than this particular woman's nose. So if we discount this body as being one of Jack the Giant Killer's victims, what could the cadaver be?

Whatever it was, it was big. And just how big? Well if you consider one of London's iconic Routemaster buses is 30 feet in length, we are looking at something measuring the length of six of them and if you have ever seen that many buses in a line that's some length of dead flesh. A couple of possible creatures from the animal kingdom have been suggested. The two candidates that spring to mind are a truly massive squid, one it has to be said that far exceeds the length of anything known today, the other is the remains of some large whale. That said even the largest known blue whale only measured 110 feet. It has to be accepted therefore that without any further clue given only the short passage in the Annals of Inisfallen and the gulf in time since the creature washed ashore that the big woman of Alba's beach is set to remain a mystery.

EDITOR'S NOTE: 'Alba' is an archaic name for Scotland, so it is impossible to pinpoint the location closely enough to provide a location map.

Stronsa Monster 1808

The next set of remains of a marine monster to wash up on the shores of Scotland was, at the time of its stranding, perhaps one of the most famous sea serpents of its day, its name: the Stronsa Monster. The remains of this creature offered perhaps the best proof of the existence of the great sea serpents of the northern seas.

It is perhaps important if we first consider the location of the sighting. The isle of Stronsa, known today as Stronsay, is the seventh largest of the Orkney islands and is some 13 square miles of low lying island, its high point being just 144 feet. It has a population of around 400, down from a total of 1,000 in the 1930s when it was supported by the herring fishing industry. But it was an altogether bigger fish that washed up in 1808.

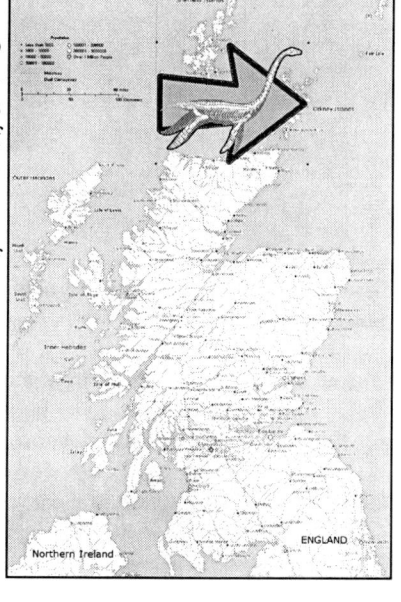

There have been many remains washed up on the world's beaches; usually these are easy enough to identify, for instance a fresh whale body is easily recognisable. Some other remains are more of a mystery, for example the St Augustine Monster washed up in Florida in 1896 was for many years believed to be the remains of a giant octopus, but later testing has revealed it to be most likely nothing more than a very impressive lump of whale blubber. Of course the Stronsa Monster was found far earlier than that and in many ways the earlier date makes the remains all the more mysterious.

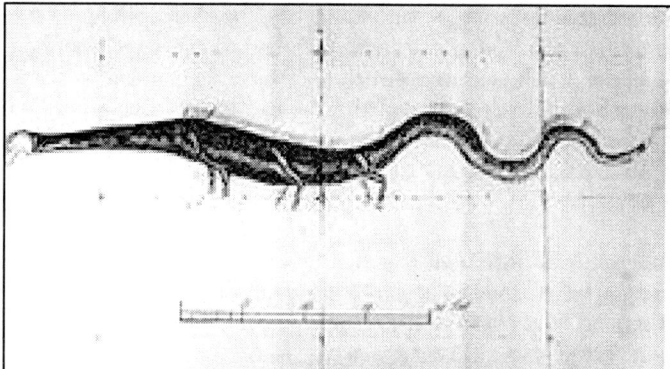

Despite the early date there was actually a great deal recorded about the creature, most of which has made its way down to the current day.

Two contemporary (1808) images of the Stronsa Monster

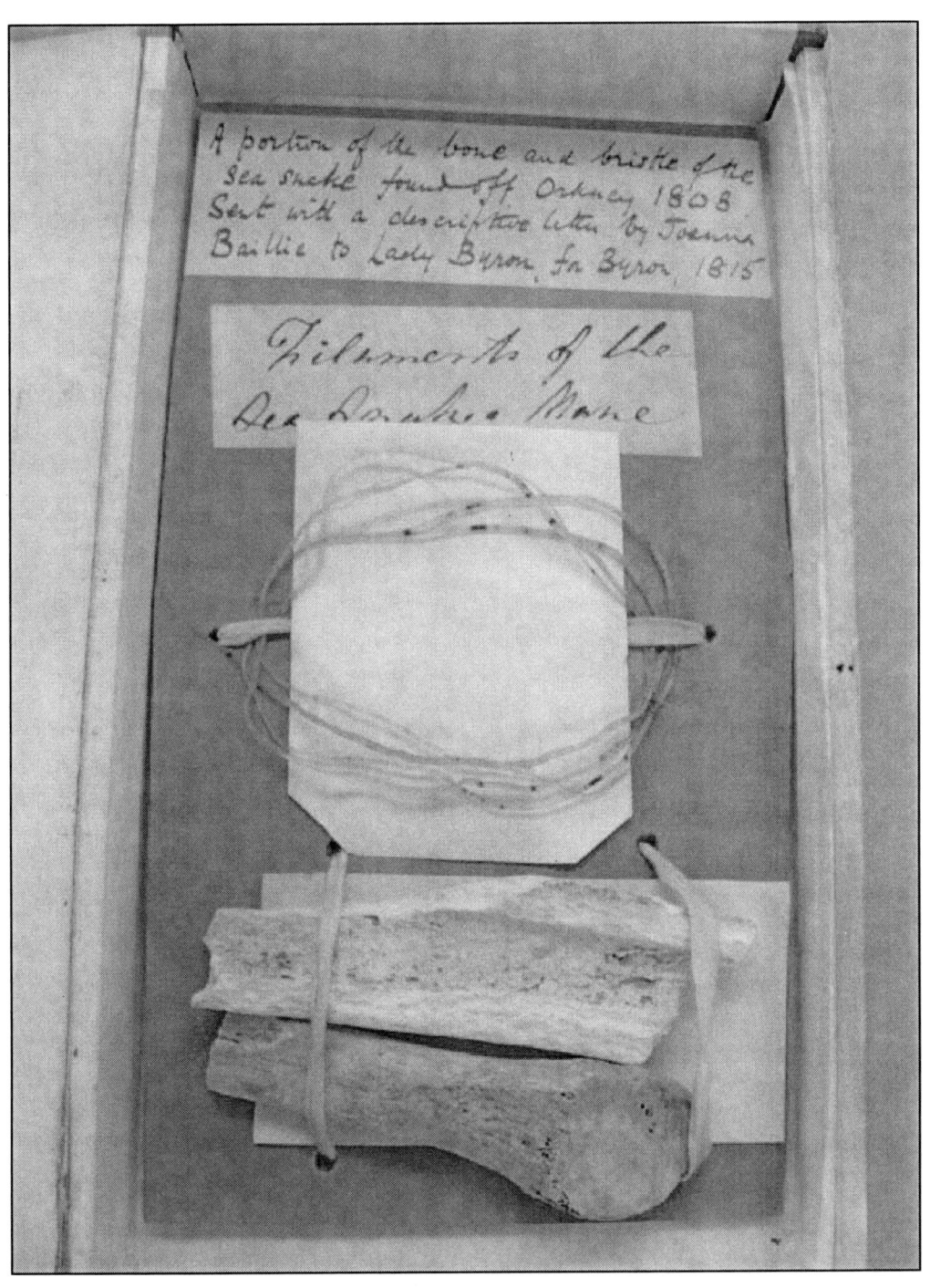

Stronsa 1808 portion of the 'Sea Snake' sent to Lady Byron

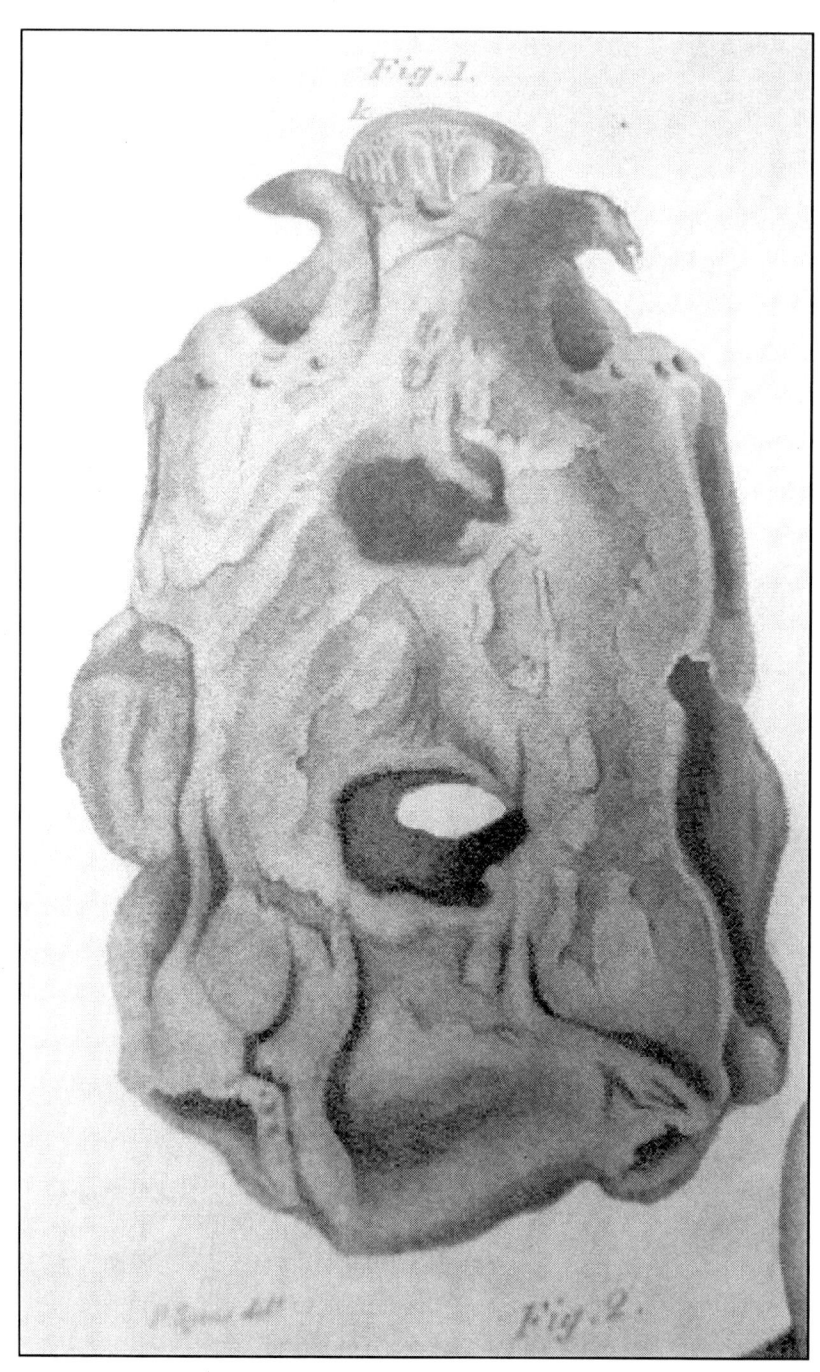

Stronsa skull 1808

Sea Serpent Carcasses—Scotland

Perhaps the best record is that noted by the Wernerian Natural History Society. This group was formed in January 1808 with the purpose of promoting the study of Sciences and of Natural History and was named in honour of the German geologist Abraham Gottlob Werner who was a creator of Neptunism. In many ways the group arrived just in time.

However the Wernerian Natural History Society is not the only source of information about the sighting. The Scottish poet Thomas Campbell who, in a great bit of name dropping, wrote to a friend on the 13th February 1809,

> 'A snake (my friend Telford received a drawing of it) has been found thrown on the Orkney Isles, a sea-snake with a mane like a horse, 4 feet thick and 55 feet long, this is seriously true. Malcolm Laing, the historian saw it, and sent a drawing of it to my friend.'

Telford was no less than the Thomas Telford, famous engineer, builder of bridges of repute, canals and roads aplenty.

But back to the Wernerian Natural History Society. It deserves the credit for the majority of the reports because they chronicled a fair bit of the information, so let's take a look at what they recorded.

Contained in the Proceedings of the Meeting of the Wernerian Natural History Society on the 19th of November, 1808 there is the following recorded,

> 'At this meeting Mr P. Neill read an account of a great Sea-Snake, lately cast ashore in Orkney. This curious animal, it appears, was stranded in Rothiesholm Bay, in the Island of Stronsa. Malcolm Laing, Esq., M. P. being in Orkney at the time, communicated the circumstance to his brother, Gilbert Laing Esq., advocate at Edinburgh, on whose property the animal had been cast. Through this authentic channel Mr Neill received his information. The body measured 55 feet in length, and the circumference of the thickest part might be equal to the girth of an Orkney pony. The head was not larger than that of a seal, and was furnished with two blow holes.
>
> 'From the back a number of filaments (resembling in texture the fishing-tackle known by the name of silk-worm gut) hung down like a mane. On each side of the body were three large fins, shaped like paws, and jointed. The body was unluckily knocked to pieces by a tempest; but the fragments have been collected by Mr Laing, and are to be transmitted to the Museum at Edinburgh. Mr Neill concluded with remarking, that no doubt could be entertained that this was the kind of animal described by Ramus, Egede, and Pontoppidan, but which scientific and systematic naturalists had hitherto rejected as spurious and ideal.'

For those who haven't a clue who Messrs Ramus, Egede and Pontoppidan are, a brief introduction may be in order. First we have Jonas Danilssonn Ramus (1649–1718), a Norwegian priest and historian, who had recorded tales of great sea serpents in his work *Norges Beskrivelse*. Han Poulsen Egede (January 31st, 1686 – November 5th, 1758) was a Dano-Norwegian Lutheran missionary who, among other things, went to Greenland looking for survivors of the Viking colonization. Not finding any he instead became a missionary to the Inuit of Greenland and was responsible for translating Christian texts into Inuit which does give us this classic from the Lord's Prayer 'Give us this day our daily harbour seal.' But it wasn't his creative use of language that found him quoted, but rather his recording of another great sea serpent of the world's oceans. This one was spotted on 6th July 1734 off the coast of Greenland, which he describes as follows,

> '...saw a most terrible creature, resembling nothing they saw before. The monster lifted its head so high that it seemed to be higher than the crow's nest on the mainmast. The head was small and the body short and wrinkled. The unknown creature was using giant fins which propelled it through the water. Later the sailors saw its tail as well. The monster was longer than our whole ship'

A very impressive sighting but these days it's often suggested that he witnessed nothing more than a very excited male whale. Putting it politely it wasn't a gun in the whale's pocket, it really was very pleased to see him.

The final member of that illustrious trio was Erik Pontoppidan, a Danish author, bishop, historian and antiquarian (busy fellow that one). In his great work *The Natural History of Norway* (1752-3) he introduced the wider world to stories of the kraken.

As you might have gathered, at the time it was understood that there was indeed a basis to these three men's writings, and the Stronsa Monster offered tantalizing proof that such a creature existed.

Back once more to the minutes from the meeting dated 14th January 1809. There is more to be said on the subject,

> 'Dr. John Barclay communicated some highly curious observations which he had made on the caudal vertebrae of the Great Sea-Snake, (formerly mentioned) which exhibit in their structure some beautiful provisions of Nature, not hitherto observed in the vertebrae of any other animal.

> 'And Mr Patric Neill read an ample and interesting account of this new animal, collected from different sources, especially letters of undoubted authority, which he had received from the Orkneys. He

OPPOSITE: Stronsa skull and paws 1808.

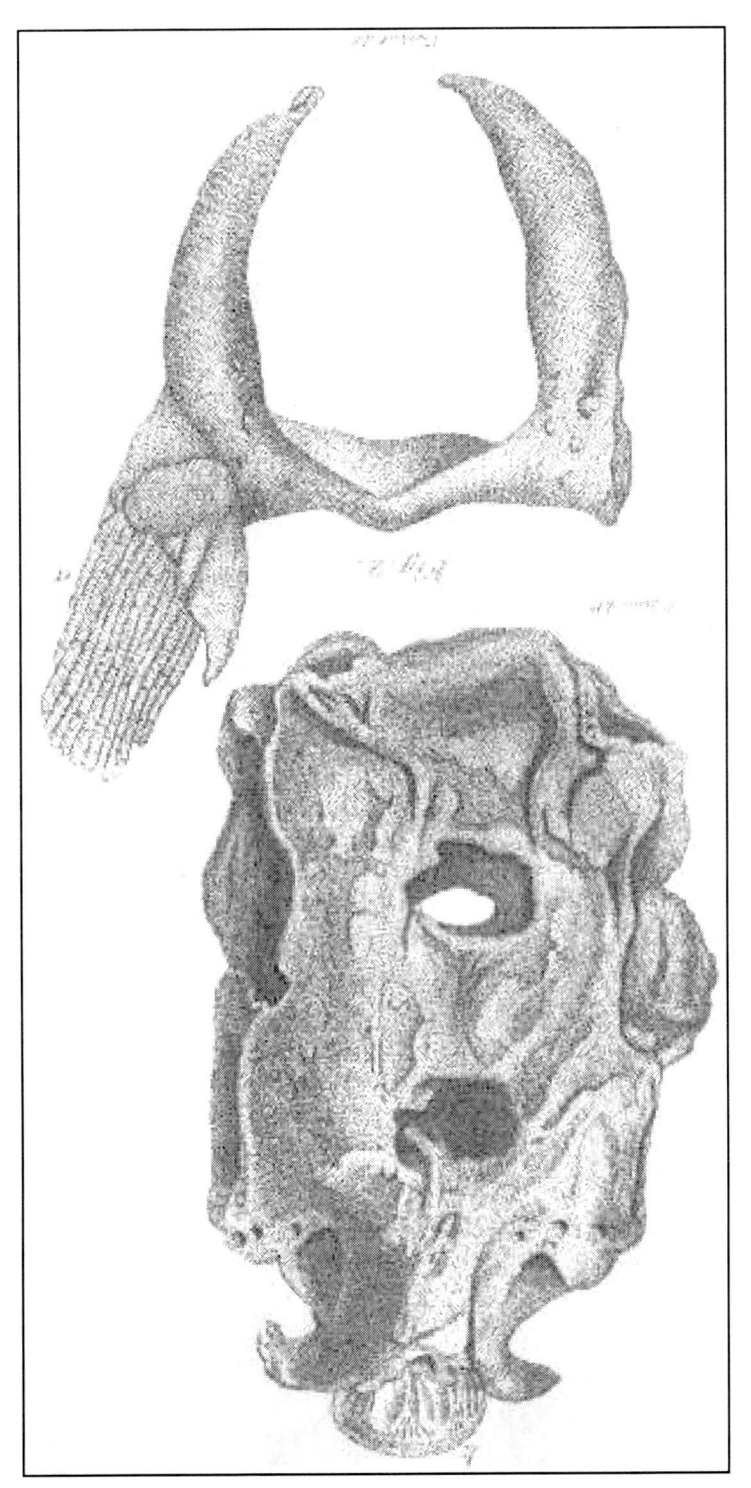

OPPOSITE: Stronsa vertebrae in glass jar.

stated, however, that owing to the tempestuous season, the head, fin, sternum, and dorsal vertebrae, promised some weeks ago to the University Museum at Edinburgh, had not yet arrived; but that he had received a note from Gilbert Meason, esq., (the gentleman on whose estate in Stronsa the sea-snake was cast,) intimating that they might be expected by the earliest arrivals from Orkney. In the mean time, he submitted to the Society the first sketch of a generic character. The name proposed for this new genus was Halsydrus, (from Üëò the sea, and ýãñïò a water snake); and as it evidently appeared to be the Soe-Ormen described above half a century ago, by Pontoppidan, in his Natural History of Norway, it was suggested that its specific name should be H. Pontoppidani.'

With that it was looking good that the Stronsa Monster really was going to be evidence of the much vaunted sea serpent; but would you believe in an animal going by the catchy name of *Halsydrus pontoppidani*? Perhaps it's not that surprising that a society such as the Wernerian Natural History Society should try and seek further confirmation of what exactly had been washed up that stormy night. With that in mind Mr Malcolm Laing and Dr Grant, living on Stronsa, were requested to take down the affidavits of the eye-witnesses. These were then presented to the Society on 11th February 1809,

'the Secretary (Mr P. Neill) laid before the Society copies of those affidavits made before the justices of peace at Kirkwall in Orkney, by several persons who saw and examined the carcass of the great sea snake (Halsydrus Pontoppidani) cast ashore in Stronsa in October last; with remarks illustrative of the meaning of some passages in these affidavits.'

And here are those same affidavits that no doubt had been awaited with a degree of excitement by the society members.

'At Kirkwall, Nov. 10. 1808. In presence of Dr. Robert Groat, Physician in Kirkwall, and Malcolm Laing, Esq. M. P. Two of his Majesty's Justices of the Peace of the County of Orkney. Compeared Thomas Fotheringhame, house-carpenter in Kirkwall; who solemnly declared, That being in Stronsa during the gales of wind in October last; he went to see the strange fish that was driven ashore in Rothiesholm Bay:

'That he measured his length with a foot-rule, which was exactly fifty-five feet, from the junction of the head and neck, where there was the appearance of an ear, to the tail: That the length of the neck, from the ear to the shoulder, was ten feet three inches, as nearly as he recollects. And being shewn a drawing of the animal, he declared, that the neck appeared to him to be too long. That the fins or arms, or, as they were called on the island, the wings of the animal, were jointed to the body

nearer the ridge of the back than they appear in the drawing:

'That the toes were less spread out, and tapering more to a point, unless when purposely lifted up; but were not webbed unless the space of an inch and a half in breadth, where they joined each other; and the length seemed to be about eight inches: That he measured one of the wings next the head, which was four feet and a half in length, and in shape, from the first joint to the extremity, it resembled a goose-wing without the feathers: That the hollow between the snout and the upper part of the skull, appeared to him not to be quite so deep as represented in the drawing: That in every other respect the drawing appears to be so exact, that if the fish had not been mentioned, it would have brought it to his recollection: That from the ridge of the back to the belly, the body appeared to be four feet in depth, and the circumference rather oval than round; but that he did not measure either:

'That the mane or bristles of the back extended from the shoulder to within two feet and a half of the tail, and were of a shining appearance when wet; but shrunk up, and turned yellow, when dried: That the mane was thin, about two inches and a half in breadth towards the shoulder, and two inches in breadth at the tail: That the skin seemed to be elastic when compressed, and of a greyish colour, without any scales: it was rough to the feeling, on drawing the hand over it, towards the head; but was smooth as velvet when the hand was drawn towards the tail: That the extremity of the tail was about two inches in thickness, and somewhat rounded; and as he saw no part of the bones, he cannot say whether any part of the tail had been broken off or not: That the eyes appeared to be no larger than those of a seal: That there were two spout holes on each side of the neck, about an inch and a fourth in diameter, and at the same distance from the head as appears in the drawing:

'That he lifted up the snout, and examined the throat, which was too narrow to admit his hand: That a part of the bones of the lower jaw, resembling those of a dog, were remaining at that time, with some appearance of teeth, which were soft, and could be bent by the strength of the hand: That he observed no nipples, or organs of generation; the belly having been burst open by the violence of the sea: That the stomach was about the size of a ten gallon cask; and the bowls about the bulk of those of a cow: That the bristles of the back which had been pulled off through curiosity, were luminous in the dark, while they continued wet. And all this he declares to be truth, &c.

'(Signed)'
'Thomas Fotheringhame.'

As descriptions go it wasn't a bad opening gambit, there is of course more. Actually there is plenty more, so here it is in all its gory details.

'Kirkwall, Nov. 19, 1808.
'Compeared John Peace, tenant in Dounatoun in Rothiesholm; and being interrogated, solemnly declares, That on the 26th. day of September last, he went a fishing off the east part of Rothiesholm-head, when he perceived as he imagined, a dead whale, on some sunk rocks, about a quarter of a mile from the Head: That his attention was first directed to it by the sea-fowl screaming and flocking about it; and on approach of it, in his boat, he found the middle part of it above the surface of the water: That he then observed it to be different from a whale, particularly in having fins or arms, one of which he raised with his boat-hook above the surface of the water: That this was one of the arms next the head, which was larger and broader than the others nearer the tail; and at that time the fin or arm was edged all around, from the body to the extremity of the toes, with a row of bristles about ten inches long, some of which he pulled off, and examined in the boat: That about ten days afterwards, a gale of south east wind came on, and the surge drove the fish ashore on Rothiesholm-Head:

That he measured it by fathoms, and found it about fifty-four or fifty-five feet in length: That he observed the six arms, or wings as they are called on the island; but perceived no part of the bristles then round the edges of the fins or arms, and supposes, that being in a putrid state, they had been beaten off by the sea, or washed away: That a small part of the belly was broken up when he saw it then, from which the stomach, as he now supposes it to have been, had fallen out: That the stomach, which he took at first for the penis, from the one end of it being joined to the body; but on seeing it after it was opened, he concluded it to have been the stomach, as it resembled the second stomach of a cow: That he did not measure the circumference of the animal, but it appeared to be of the thickness of a middle sized horse round the girth, of twelve or thirteen hands high. And being shewn a drawing of the animal, and desired to point out the resemblance or difference, he declared, That the Joint of the foremost leg was broader than represented in the drawing, being more rounded from the body to the toes, and narrower at the upper end than at its junction with the toes: That the limb itself was larger than the hinder ones, and the uppermost joint or shoulder was altogether attached to the body: That in all other respects the drawing appears to him to be an exact resemblance of the fish, as it lay on the beach: That the mane came no further than the shoulder, and extended to the tail, part of which appeared to have been broken off: That the length of the neck, the situation of the spout-holes, and of the eye, the shape of the snout, the position and distance of the limbs from each other, appear to him to be exactly preserved in the drawing: That the lower jaw was awanting when he saw it: That the fish was of a greyish colour: That he observed no nipples or organs of generation, unless as above mentioned: That the part of the belly which was burst open, and from which the stomach had fallen out, was between the two limbs that are situated in the middle of the animal. And all this he solemnly declares to be truth. And declares he cannot write.'

'Eodem die'

'Compeared Mr. George Sherar, tacksman of Rothiesholm in the island of Stronsa; who being interrogated, solemnly declared, That on the 20th. of October, being in Rothiesholm-head he saw the crew of John Peace's boat examining something on the water, which he took to be a dead whale: That about ten days afterwards, a gale of east wind having taken place he went to see if the whale was driven ashore, and found it in a creek, lying on its back, about a foot under water; and from the view which he had of its figure, length and limbs, his curiosity induced him to return a day or two after the gale had abated, when he found it thrown upon the beach, a little below high water mark, and lying on its belly, as represented in the drawing: That he returned next morning, with a foot-rule, purposely to measure it, and found it to be exactly fifty-five feet in length, from the hole in the top of the skull (which he has brought to town with him), to the extremity of the tail: That the length of the neck was exactly fifteen feet, from the same hole to the beginning of the mane: That he measured also the circumference of the animal as accurately as he could, which was about ten feet, more or less; and the whole body, where the limbs were attached to it, was about the same circumference: That the lower jaw or mouth was awanting; but there were some substances or bones of the jaw remaining; when he first examined it, which are now away: That it had two holes on each side of the neck, besides the one on the back of the skull: That the mane or bristles were about fourteen inches in length each, of a silvery colour, and particularly luminous in the dark, before they were dried: That the upper part of the limbs, which answers to the shoulder-blade, was joined to the body like the shoulder-blade of a cow, forming a part of the side: That a part of the tail was awanting, being incidentally broken off at the extremity; where the last joint of it was bare, was an inch and a half in breadth: That the bones were of a gristly nature, like those of a halibut, the back-bone excepted, which was the only solid one in the body: That the tail was quite flexible, turning in every direction, as he lifted it; and he supposes the neck to have been equally so, from its appearance at the time: That he has brought in, to deliver to Mr. Laing, the skull, two joints of one of the largest limbs, next the head, with different parts of the backbone, besides the bones that were formerly sent in: That there were either five or six toes upon each paw, about nine inches long, and of a soft substance: That the toes were separate from each other, and not webbed, as far as he could observe; and that the paw was about half a foot each way, in length and in breadth: That a few days thereafter, a gale of wind came on, and drove it to another part of the shore, where it was broken to pieces by the surge, and when Mr. Petrie came out to take a drawing of it, no part of the body remained entire: That he endeavoured to convey an idea of the animal to Mr. Petrie, by drawing the figure of it as accurately as he could, with chalk, on the table, exactly as it lay on the shore, after which Mr. Petrie made six or seven different sketches or plans of the fish, before he could bring it to correspond, in each minute particular, with the strong idea which he retains of its appearance: That he was the more attentive to its shape, dimensions and figure, in order to be able to give an accurate account of it to any travellers that might come to Rothiesholm, and that he is ready to make oath that the drawing is an exact resemblance of the fish, as it appeared when he measured it; and corresponds in all particulars with the idea which he entertains of the figure, dimensions, and proportions of the fish: That the substance of the body appeared like coarse, ill coloured beef, interlarded with fat or tallow, without the least resemblance or affinity to fish; but when put into a lamp, and the lamp placed on the fire, it neither flamed nor melted, but burned away like a gristly substance: That he perceived no teeth in the upper jaw; the lower jaw and tongue

being awanting, and the palate also away: That the aperture of the throat appeared to be so wide, that he might have put his foot down through it: That the joints of the limbs were not united by a ball and socket but were lapped over each other, and united by some means which he does not comprehend That there were two canals, one above and another below the backbone, large enough to admit one's finger, and extending from the vertebrae of the neck, to the extremity of the tail, containing two ligaments, which he supposed, enabled the animal to raise itself up, or to bend its body in a spiral form: That a tract of strong easterly wind had prevailed, before the body was discovered upon the shore, and that he saw the body on two or three different occasions, after he had measured it, and before it went to pieces.
'And all this he declares to be truth, &c.'
'(Signed)'

'Geo. Sherar.'

What's that I hear you say? You want more? Well you're in luck, there is some more.

'Compeared Mr. William Folsatter, tacksman of Whitehall, in the island of Stronsa; who being interrogated, solemnly declared, That having heard that it was a dead whale that had come on shore in Rothiesholm-head, he did not see the body till about the 28th. day of October, when it had gone to pieces: That he saw about nine or ten feet of the back-bone, and some bones of the paws, and what was supposed to be the stomach which last he had the curiosity to open; that it was about four feet long, and as thick as a firkin, but flatter: That the membranes that formed the divisions, extended quite across the supposed stomach, and were about three sixteenth of an inch in thickness, and at the same distance from each other, and of the same substance, with the stomach itself: That the section of the stomach, after it was opened, had the appearance of a weaver's reed: That he opened about a fourth part of the supposed stomach which contained nothing but a reddish substance, like blood and water, and emitted a fetid smell: That he was very doubtful at the time whether it was really the stomach or not; but that each end of it had the appearance of terminating in a gut. And all this he solemnly declares to be the truth, &c.'
'(Signed)'
'Wm. Folsetter.'

Still with us then, there's just a little bit more from our witnesses.

'The said Mr. George Sherar being again interrogated, declares, That he examined the supposed stomach, after it had been opened by Mr. Folsitter, and that he laid it open to the farther end: That there was something like a gut at the end which he opened, about two inches long, with a small aperture: That the stomach had the same appearance from end to end, and contained nothing but a substance like blood and water: That the large bone of which a drawing was taken, was considered as the collar-bone; and that it was situated with the broad and thick part downwards and the open part towards the vertebrae of the back: That he observed no appearance of fins about the neck or breast, or other parts of the body, except the six paws already described. And all this he solemnly declares to be truth, &c.'

'(Signed)'

'Geo. Sherar.'

Well if nothing else, we have to acknowledge that this stranding was well recorded, and it would appear for the time a very good deal of effort went into the study of this mystery animal. But was this great carcass really proof of the existence of the great sea serpent, or was it just the body of something much more mundane that had washed up? The next person to wade into these waters was Everard Home.

Now as you may expect there were those then as there are today who look at the evidence and see something completely different, and Everard Home was one such man. A noted ichthyologist his verdict was that the Stronsa Monster was nothing more than an unfortunate basking shark that had come to a sticky end somewhere off the coast of Orkney.

His findings were noted in the *Philosophical Transactions of the Royal Society of London*, Vol. 98. Rather lovingly entitled 'An anatomical account of the *Squalus maximus* (of Linnaeus), which, &c'

With a title like that you expect something good to follow, and his comments certainly give pause for thought when considered recent strandings of large basking sharks,

> 'especially of an individual of thirty feet six inches, "entangled in the herring nets, belonging to the fishermen of Hastings, 13th Nov. 1808".

> 'I cannot close the present paper without mentioning, that nearly the same period, two other Squali of large dimensions were thrown upon our coast. The probable cause of this event, is the season being uncommonly boisterous and tempestuous. On the 3rd. Of January, 1808, a fish was thrown ashore at Penrhyn, in Cornwall. On hearing of it from a person on the spot, I sent down a drawing of the subject of this paper to compare with it, and the fish proves to be of the same species, and a male, measuring thirty-one feet in length.'

> 'The other was thrown ashore on the 7th. of October, 1808, at Rothiesholm, an estate of Gilbert Meason, Esq. in Stronsay, one of the Orkney isles. It had been seen lying on some sunken rocks, eleven days before, was in a half putrid state, and the sea fowls were in great numbers feeding upon it. Those who saw it, reported that the skin was rough in one direction, and smooth like satin in the other. At the time of its

Stronsa Monster, original oil painting by Glen Vaudrey
Oil on stretched canvas measuring 16 x12 inches (40cm x 30cm)

Capturing the monster in all its glory this painting shows it as it may have looked when freshly washed up, the Highland gent looking on gives an idea of the size of the beast.

being examined, the skin and a great many other parts of the fish were wanting.'

'Mr. Meason, with a zeal for science which does him infinite credit, upon hearing the strange accounts which were given of this sea-monster, got his brother, Malcolm Laing, Esq. and Dr. Grant, an eminent physician (both justices of the peace), to take depositions on the spot, from those persons who had seen the fish, that its real appearance might be ascertained. This examination, however, did not take place till six weeks after the fish was thrown ashore.'

'These depositions were sent to Sir Joseph Banks, who put them into my hands. (The depositions are very long, and exceedingly minute; they are preserved in the Board-book of the Royal Society). I also received, a short time after, from my friend Mr. Laing, in consequence of a request I made for that purpose, that part of the skull, which contained the brain, the upper jaw having been separated from it, a considerable number of the vertebrae of the back united together by their natural attachments, a portion of one of the pectoral fins, with the cartilages that unite it to the spine, and a long and short cartilage forming the support of one of the gills. On comparing these different parts, with those of the *Squalus maximus*, they were found to agree, not only in their form, but also in their dimensions. This led to the opinion of the fish being a Squalus, a very different one from what was formed by those who saw it in the mutilated state in which it was thrown ashore, and who called it a sea-snake. In the different depositions, several parts are accurately described, such as the valvular intestine, which was taken for the stomach, and the bristles of the mane, which are described as ligamentous fibres, one of them is in my possession, and is of the same kind with the fibres forming the margin of the fins of the *Squalus maximus*. The drawing that was made from memory, and which I have annexed, will enable me in a few words to point out how much, in some things, those who saw the fish adhered to truth, and in others allowed their imagination to supply deficiencies, for one of them declared, with confidence, that the drawing was so exact a representation of what he had seen, "that he fancied he saw the beast lying before him, at a distance on the beach."'

'The drawing is correct in the representation of the head, and anterior part of the fish, from which the skin, the upper and lower jaw, the gills, and gullet, had been separated by putrification; and when we consider that the liver and the other viscera were all destroyed, except the valvular intestine, which was taken away by the observers, the size of the body that remained would be nearly in proportion with the drawing. The legs are tolerably exact representations of the holders in the male *Squalus maximus*, described in a former part of this paper, and therefore are not imaginary, only that four have been added which did not exist. This is satisfactorily determined by the pectoral fin, which is preserved, having no resemblance to them. The mane, they said, was composed of ligamentous fibres, one of which was sent to London; this corresponds, in its appearance, with the fibres that form the termination of the fins and tail of the *Squalus maximus*, such an appearance therefore was seen, but could only be met with in the place of the two dorsal fins, instead of being continued along the back, as in the drawing. The contortions towards the tail are such, as the invertebral joints could not admit of, they are therefore imaginary.'

'It is said, two different persons measured the fish; one by fathoms, the other by a foot-rule, and that it was fifty-five feet long. Their accuracy is at least doubtful, as the parts that are preserved correspond with those of a fish about thirty feet long, and it is rendered still more so, as the person who gives the length in fathoms, says, he saw at that time the six legs, the two foremost being larger than the hinder ones, and the lower joint more rounded from the body to the toes. The pectoral fin, which is preserved, proves this declaration to be incorrect: the person who measured the fish with a foot-rule, declares the length, from the hole in the head to the beginning of the mane, to be exactly fifteen feet, which is probably correct since a Squalus of about thirty-six feet long would measure, from the forepart of the skull to the dorsal fin, about fifteen feet; but the other measurement must be questionable.'

'It is deserving of remark, that there is no one structure represented in this drawing, which was not actually seen. The skeleton of the holders corresponds with the legs in the drawing, the margin of the dorsal fin in a putrid state with the mane; so that the only errors are in the contortions towards the tail, the length of the fish and the number of the holders, which were mistaken for legs. (This mistake of the holders of the male shark for legs, has been frequently made. There is a drawing in Sir Joseph Bank's library, sent from Ireland, in which the fish is represented walking like a duck, with broad webbed feet. The skin of a male *Squalus maximus* was exhibited in London, some years ago, distended by means of hoops, and the holders were shown as its legs, on which it occasionally walked). And when we recollect that the drawing was made from memory six weeks after the fish had been seen by those, who describe it, during which interval it had been their principal subject of conversation, we may conclude that so extraordinary an object, as the mutilated fish must appear, when believed to be a perfect one, would, in their different discourses, have every part exaggerated, and it is only remarkable that the depositions kept so close to the truth as they have done.'

'It is of importance to science; that it should be ascertained, that this fish is not a new animal unlike any of the ordinary productions of nature, and we are indebted to the zeal and liberality of Mr. Meason and Mr.

> Laing, who have collected a sufficient body of evidence. to enable me to determine that point, and prove it to be a Squalus, and the orifices behind the eye, which communicate with the mouth met with in the skull, renders it very probable, that it is a *Squalus maximus*.'
>
> 'This opinion is further confirmed by the *Squalus maximus*, known by the name of the basking shark, being frequently seen upon the coast of Scotland.'

Now there are many possible things that you may think of Everard Home's findings. I, for example, do have to consider that he might be on to something in suggesting that the Stronsa Monster was nothing more than a very rotten shark that had been beaten around by the actions of the sea much to its detriment. Certainly some of the features that make the creature a sea serpent are also to be found in the shape that a rotting basking shark goes for. You have to admit his suggestion as to what the six legs are sounds more likely than the idea that there is some great creature in the oceans with six legs. Certainly the basking shark could have all the right pieces to look the part to anyone unfamiliar with it, and they could certainly be suggestive enough to be mistaken at a casual glance.

There is one remark however that I have to point out about Mr Home's findings and that is the belief that the measurements are wrong. While it's possible for witnesses to be mistaken in the identification of a rotting mass of flesh, it's asking a little too much to then suggest that the many gathered witnesses were unable to take a true measurement. It's fair to say that they could be mistaken in measuring a few feet, perhaps it was a foot or two either side of 55 feet, but could you really believe someone being 20 foot out, that's a big difference. 'And just how big a difference?' I hear you ask. Think the size of a dry cargo shipping container. Is it really possible for more than one person to be that far out?

It's often the case unfortunately, that evidence that, by its nature, questions a strongly held belief will be ignored. Home's basking shark theory is let down by the beast's length. The great serpent idea also has its failing, for example, in not considering the idea that it could be nothing more than shark remains. There is of course nothing especially unusual about such handling of data, it was this kind of damned data that would give the title to Charles Fort's book *The Book of the Damned*.

Now you will hardly be surprised that the previously mentioned Dr Barclay was not all that impressed with the findings of Mr Home, and was so moved as to write a paper on the subject, which - fortunately for the modern reader - was published in the first volume of the *Memoirs of the Wernerian Society*. So let's have a peep at what he had to say about it.

> 'Since reading the first paper of Mr. Home, where he treats of the vertebrae of the *Squalus maximus*, I have seen another, entitled 'An anatomical account of the *Squalus maximus*'. In this last paper, he seems to be convinced, that the animal of Stronsa is a *Squalus maximus*. The scale on which he draws his figure of the squalus, is a scale of half an inch to a foot.

'Measuring by this scale, the head of his squalus is five feet and a half, from the joint of the upper jaw to the gills. The dried and shriveled head of the animal of Stronsa, measures only twelve inches from the first vertebra to the farthest part that remains of the jaw.

'The diameter of the head of the *Squalus maximus*, from right to left, at the angle of the mouth, was, according to Mr. Home, five feet. The broadest part of the head of the animal of Stronsa is, in its present state, only seven inches.

'The diameter of the larger vertebrae, near the head, in the squalus, was, according to Mr Home, seven inches. The first cervical vertebra in the animal of Stronsa, is still adhering to the head, and is only two inches in diameter.'

'Yet some of the vertebrae of this animal, which are still preserved, are six inches and a half in diameter; and the first vertebrae which I saw, were from four to five and a half inches across.

'The smallness of the cervical vertebrae, in the animal of Stronsa, confirms the account of those who saw it, that the animal had a neck. But the *Squalus maximus*, if Mr. Home's figure be accurate, had nothing resembling a neck. And, indeed, Artedi observes, that "*omnes pisces qui pulmonibus destituuntur, collo quoque carent: Ergo soli pisces cetacei collum habent.*" The presence of a neck, therefore, as peculiar to cetaceous fishes, confirms likewise the account of the spiracula or ear-holes, ascribed to this animal of Stronsa.

'The length of Mr. Home's squalus was thirty feet six inches. The length of the animal of Stronsa, by actual measurement was fifty five feet, or, exclusive of the head, fifty four; and yet a part of the tail was supposed to be wanting. The circumference of the animal of Stronsa was, by actual measurement, about ten feet, meant, I suppose, at the thickest part. If the animal had been cylindrical at that part, the diameter from the dorsal to the sternal aspect must have been about three feet four inches. The diameter of the squalus at the thickest part, measuring from the dorsal to the sternal aspect, is nearly six feet; its circumference, had it been cylindrical nearly eighteen feet.

'The animal of Stronsa had a mane, extending from the shoulder to near the caudal extremity (i.e. about thirty nine feet), after deducting the length of the head and neck, which, when together were sixteen feet. I have still a specimen of that mane, which I got from Mr. Urquhart; and all the specimens which were brought here, confirm the accounts that were sent of it from the Orkneys. The bristles of that mane are not like the radii of a fin, nor, although they were, has the squalus a fin extending from the shoulder to the tail.

'A drawing, which was sent to me by our very active and obliging Secretary, Mr. Neill, was executed, I am told, from the original, by Mr. Urquhart; and its accuracy is confirmed by the dried specimen now before us. It represents the sternum and two parts corresponding two scapulae, and those organs which are named paws. Mr. Home says, that these organs resemble the pectoral fins of his squalus. But the length of the pectoral fins, measuring along the upper margin, is four feet, the length of the paw cannot be determined, as part of it is wanting; the part that remains, measures seventeen inches.

'The breadth of the fin, measuring across the radii, is three feet and seven inches; while the greatest breadth of the paw in its dried state, is only five inches and three quarters.

'Those parts which in form resemble the scapulae and exhibit articular surfaces at each extremity, were probably ribs.'

Dr Barclay concludes by observing that,

'that "it is of importance to science, that it should be ascertained, that this fish is not a new animal, unlike any of the ordinary productions of nature." Of what importance it is to science to admit no new genera or species into our catalogues of natural history, I cannot conceive. But it is certainly of much importance to science, that the naturalist should be cautious not to determine the species of an animal upon vague evidence. Now what evidence had Mr. Home that this animal was a squalus, and even to suppose that it was a *Squalus maximus*?'

It's starting to take on the look of theory tennis, both sides seem happy to be batting their argument across to the opposition without any thought or consideration to what the other side is actually saying other than to knock the stuffing out of it. In Dr Barclay's response we have something else to ponder, the two men comparing measurements, in this case Dr Barclay doesn't seem to take into account the effects of a rough sea and general degradation of the Stronsa Monster. It would be fair to say that the remains of the creature were well past their best by the time they were first observed. It was of course this damage that has made the remains such a mystery, the corpse providing no clear indication of the real nature of the unfortunate animal, just a hint of sea serpent about it as well as a hint of basking shark.

While it would appear that Dr Barclay was indeed on to a loser with the beast being a sea serpent it was far from certain that Mr Home had the answer, and there were more people who would chip in with an idea or two, and I don't see any reason to miss any of those ideas out.

A certain Professor Lorenz Oken, a German natural philosopher and editor of the *Isis*, waded into the debate in 1811. He added his own take of what the creature could have been. After stating that no animal that has a bony skeleton has six feet, he therefore noted that it must be a cartilaginous fish, and a male one at that for the last two legs are actually most likely to be nothing more than two pterygopodia (a pair of additional paring-organs, the so-called 'claspers' or 'holders'). These claspers, when added to the ventral and pectoral fins, provide you with just enough feet for the Stronsa Monster. Now Professor Oken decided that the Stronsa Monster was neither a shark nor a cetacean. No, he opts for something completely different as a likely suspect, but let's see how he put it,

> 'The animal consequently is more related to the sharks, and as it is not a true shark, it must be a Chimaera.'

I know what you are thinking, no description of the Stronsa Monster even vaguely sounds like that terror of the ancient world, the Chimaera, an unholy hybrid of lion, goat and snake. Well you would be right, this was a completely different Chimaera, this one is the *Chimaera monstrosa*, a deep sea fish. It appears that Professor Oken was taken with this fish as being the likely candidate due to its possession of a dorsal crest that runs from just behind the head all the way along to the tip of its long thin tail.

Just in case you think that Professor Oken could have been onto something he goes and shoots himself in the foot with both barrels by adding the following,

> 'finally individuals of Chimaera of 30 feet in length, have already been caught.'

The trouble with that statement is that he hasn't quite got the length correct - actually he hasn't got the length correct in two ways. First of all the widely accepted measurement of the Stronsa Monster is some 55 feet not 30 feet, but that's nothing compared to the real clanger. You see they just don't catch Chimaera anywhere near the length of 30 feet, in fact the longest recorded specimens have not even reached 5 feet, well short of 30 feet, and so far removed from 55 feet as to really stretch the imagination. Perhaps then, in the quest for the identity of the real Stronsa Monster, we should consider the prof's ideas as interesting but nothing more.

Of course there have been plenty of others who, having considered what those rotting remains could possibly be, have come up with some very interesting ideas. It does have to be said that they are largely based on their own desire to show that they have found proof of some theory or other. So with the caveat of 'don't believe everything you read', let's have a look at some of these.

When, in 1822, Dr Samuel Hibbert, a Mancunian with an interest in Shetland, published his rather catchily titled *A Description of the Shetland Islands: comprising an account of their geology, scenery, antiquities and superstitions*, he included his own thoughts on what was represented by those remains washed up on the beach, and had no doubt in his mind as to what they were when he stated,

> 'The existence of the sea-snake,—a monster of fifty-five feet long, is

> placed beyond a doubt, by the animal that was thrown on shore in Orkney, the vertebrae of which are to be seen in the Edinburgh Museum.'

Sadly the giant sea snake is still awaiting discovery, but who knows one day it might wash up on a lonely beach. Absolutely nothing to do with mystery animals, but I thought it interesting to learn that Dr Hibbert ended up, like me, living in Cheshire, however he ended up on the family estate. Short of getting a car with a very large boot, I doubt the same could apply to me.

The next person to cast a critical eye over the remains was Dr Robert Hamilton who also opted for it being a sea serpent. He considered the carcass in his book *The Natural History of the Amphibious Carnivora; Including the Walrus and Seals, Also of the Herbivorous Cetacea, &C.*[*] In 1839 it was already some 21 years after the carcass had washed up and the smell might well have gone, but interest in it still remained strong. When considering the case he starts,

> 'We turn, first, to an account of an animal which apparently belonged to this class, which was stranded in the Island of Stronsa, one of the Orkneys, in the year 1808.'

With such a promising start he then goes on to look at the evidence before giving his verdict at the end of the page,

> 'many affidavits proffered by most respectable individuals as well as from other circumstances narrated leaves no manner of doubt as to the existence of some such animal.'

Again we might have to think about that one. He might not have been entirely correct, still no harm in trying.

On to the next person to have a dabble at second guessing the nature of the Stronsa Monster's remains. This time we are off to Germany and Professor Dr W.F. Erichson, the editor of the *Archiv für Naturgeschicht*. He had studied the details recorded by Mr Barclay, and the descriptions of the parts of the animal that had been saved; having considered this evidence he stated,

> 'All these parts belong undoubtedly to a shark,'

So far so good, there has to be little doubt that despite the wish for the carcass to be that of a sea serpent, all the evidence does actually point to its being the remains of a shark. However, the good Professor goes a little bit further and gives us his opinion as to which shark.

> 'Everard Home already declared the animal to be a shark, and in spite of all that Dr. Barclay asserts to the contrary, it will be so for ever, only it may not have been a *Selache maxima*, but a *Lamna cornubica*, which

---

[*] There are, of course, no herbivorous Cetacea - this is an outdated term for Sirenia.

also reaches a considerable length. So the animal of Stronsa has no relation at all with the sea-serpent of the Norwegians.'

Already we know that the *Selache maxima* is the basking shark but what is the *Lamna cornubica*? Well for those who don't know, it's the porbeagle shark. Unlike the basking shark, this shark is no plankton feeder but rather an active predator, which - you will be glad to know if you are paddling in the chill waters of Orkney - isn't known for attacks on people. Of course, as with many theories concerning the Stronsa Monster, it has its flaws and I might as well go straight to the big one, and as ever with the Stronsa Monster, it is a question of length. As we know, the remains as measured were some 55 feet long. Now the porbeagle is a little tiddler by comparison, only being known to reach around 12 feet in length. Certainly a big fish and I for one wouldn't be that comfortable swimming alongside it in the local swimming pool, but it is nowhere near as long as the Stronsa Monster, which to be honest wouldn't even fit in the local pool. Had it been that the porbeagle could reach such lengths we would surely have one of the world's most fearsome predators lurking beneath the waves.

Now while the Stronsa Monster was looking to be anything but a sea serpent, the reporting of it did have some positive feedback. One such occasion was the prompting of the Reverend Donald Maclean of the small isles (Inner Hebrides) to report his own encounter of a sea serpent to the Wernerian Natural History Society, that being a sighting that he had had of a 70 foot long sea serpent off the coast of Coll in June 1808. Had it not been for the publicity surrounding the Stronsa Monster possibly his report would not have been made. Not surprisingly, this sea serpent was altogether more sea serpent-like than the corpse on the beach at Stronsa and if you would like to know more about what Reverend Maclean saw that day I would direct you towards reading *Mystery Animals of the British Isles: The Western Isles*.

Back to the Stronsa Monster. There are of course a few unanswered questions remaining, the big one being 'what was it?' And the equally important and intriguing question 'what happened to the remains?

Let's begin with the story of the remains, considering that the carcass started off as some 55 feet of rotting meat you might be a little surprised to hear that any of it was retained (think of the smell for a start), but in fact some of it was stored, and there are still a few bits to be found lurking in dark places even today. One of the first people mentioned in this tale of the strange fish, Malcolm Laing, took the very sensible action of actually collecting some samples. You would think that it would be the first thing that anyone would think of, but of course looking at the parts is nowhere near as much fun as claiming loudly that the remains are those of a true sea serpent or a badly measured small shark.

Mr Laing managed to take these samples before the weather finally did for the remains. While some of the samples were sent to Edinburgh, other parts headed off in different directions. The skull, it is reported, went off all the way down to London, where many years later it would be destroyed like so much during the Blitz.

Some remains found their way into the hands of Joanna Baille, the Scottish poet and dramatist.

She managed to get her hands on some, and would send them off to Lady Byron as she considered that it would be of interest to Lord Byron. Through the research of Dr Beale, the current location of these remains was found to be part of the John Murray Collection at the National Library of Scotland.

The other surviving pieces collected found their way into the old Natural History Museum where they were looked at by Professor Goodsir in 1849. Sometime around 1864, along with the rest of the contents of the Natural History Museum of the university, they were transferred to the Government Museum, which soon changed its name to that of the far better known Royal Scottish Museum. In the words of James Ritchie of the Natural History Department of the University of Aberdeen, they lay in obscurity until he rediscovered them and re-examined them in 1933.

From his study of these remains, Ritchie was yet another to confirm that they were indeed not those of a sea serpent or plesiosaur, but rather most certainly the remains of a gristly fish. In fact he went as far as to say that they very well matched the remains you would expect to find from a basking shark, sharing as they did the texture of segments of the vertebrae, their size and curious pillared structure. The bristles of the mane that had so excited so many in the previous century appeared to be nothing more than the frayed out tissue of a large back fin.

But what of the remains today, I hear you ask. Well the tale continues.

During my research into the Stronsa Monster I was fortunate enough to contact Dr Yvonne Beale who had been carrying out some fine research on identifying the Stronsa Monster once and for all. A geneticist, Dr Beale - who has a degree in evolutionary, environmental and biomedical genetics from the University of St Andrews and a PhD in the field of DNA damage repair - realised that in these days of DNA testing there would be a good chance that this previously unavailable avenue to researchers could well provide the answers to the big question of what the creature was. It was hoped that a small sample of material from the surviving remains of the Stronsa Monster could be sent to a DNA testing laboratory in Florida, which has a database of all known shark species, where it could be tested to see if it matched any shark. It all sounds so simple (I am sure it isn't). While it might only take the tanned folk of CSI Miami around three minutes to get a DNA record from whatever scrap they find, it has proved a little more difficult in real life. You see they don't have to contend with the National Museum of Scotland who, I have to report, refused Dr Beale's request for a sample to test. If only there was a Horatio Caine of cryptozoology, we would have the results by the end of the chapter. A very frustrating end to what promised to be a true leap forward in the tale of the Stronsa Monster.

So what is the Stronsa Monster when all is said and done? Well to be honest, there are still many theories; in some quarters it is still believed to be a sea serpent or a plesiosaur, but I think the smart money is on its being a member of the shark family. As we have seen, many a person has tried to identify the remains as one shark or another, but the main stumbling block has been the size of the carcass; at some 55 feet it is considerably larger than you would expect to find from the remains of the most likely noted suspect, the basking shark. Unlike some earlier commentators, I see no reason for deciding that the witnesses were unable to measure the remains correctly and at 55 feet it would far exceed the 33 feet you might expect a lucky, and well fed adult basking shark to reach. I say lucky because there has been considerable hunting of basking sharks since 1808, which may of course be the reason that none has been found that has managed to achieve such an impressive length due to a premature end at the end of an exploding harpoon. That's not to say that there haven't been larger basking sharks recorded; the largest accurately measured specimen was trapped in a herring net in the Bay of Fundy, Canada, in 1851, it was some 40 feet in length so still well short of the Stronsa Monster, but getting nearer.

If we assume for a moment that the real reason the Stronsa Monster is some 22 feet longer than the average basking shark is because it isn't a basking shark at all, but some other kind of shark, we then have a problem because few sharks get much bigger. Only the whale shark is known to exceed the basking shark in length and even then the recorded length for the type is still less than 44 feet. So this, added to the fact that the whale shark's habitat of choice is tropical and warm water oceans, certainly not something that can easily be equated to the waters around Orkney, doesn't give us the answer.

Now there is one shark that is estimated to fit the size, and that is the *Carcharodon megalodon*. This fearsome predator is estimated to have ranged in size (depending who you believe) anywhere from 56 feet to a really impressive 98 feet. In case you have never seen a picture of a megalodon, it would be fair to say a scaled up great white shark would give you an idea of what it looks like, certainly it would be enough to put off all but the most suicidal of swimmers from venturing into the sea. It would easily fit the size element of the mystery, but it does have a few drawbacks that should really be considered before you decide that you never want to go swimming again. And the biggest drawback? It is widely believed to have been extinct for the last 1.5 million years (but that's not to say people don't still report sightings occasionally).

Scalloway 1810

Just under 200 years ago there was something very strange to be found in the waters off Scalloway. What exactly it was is hard to say as there doesn't really seem to be all that much in the way of a description, but that's not due to people not noticing it rather than to the fact that they were doing their best to avoid it. 'What,' I hear you say, 'could possibly be in the water that would have you sticking to the shore? A giant mystery shark? A scary looking high-finned whale?' No, neither of those. It was nothing less than the fearsome kraken.

Now I bet you are wondering what a kraken actually is, and to be honest that is a very good question and one which many people have been trying to answer for many years. In Norse mythology it was a vast animal with a body that had a circumference of around 1½ miles. This, no doubt, explains why some sailors had a tendency to

think that it was an uncharted island, a mistake that could on occasions have some serious repercussions. Not only could the island sink at very short notice, but if that wasn't bad enough the kraken would also occasionally attack passing ships by grasping their rigging with its many tentacles. So far that hasn't really given any clues as to what it could actually be, and certainly there have been many candidates for the true form of the kraken, these ranging from it being some unknown but powerful whale, the great sea serpent himself, even a giant crustacean; all have been suggested for the creature's identity. However, there is one animal that seems to have become associated with the name of the kraken, and that is the giant squid. Now you may wonder if this creature of Norse legend has ever been sighted in recent years. Well the answer is no, not really, certainly not within living memory, but if you let in stories from the last couple of hundred years there certainly has been an occasion, and that is said to have taken place in the waters off Shetland.

It was in the year 1810 that something very unusual turned up in the waters near to the village of Scalloway, the one time capital of Shetland. Today if you go to Scalloway you may consider Scalloway Castle (opposite page) as the sight to see, and it is a nice place to visit, but just two centuries ago, it was the bay that had the object worth viewing because it is said that back then there was a kraken floating in the waters.

It was recorded that in the January of 1810 there was a kraken, or at the very least a monstrous fish, floating off Scalloway. This mystery creature was described as resembling some vessel turned upside down so that its keel was exposed to the air; it could, according to some, be taken for a small ridge of rock or a small island. So strange was this creature that none from the island would travel out to it to see what it could be.

Now you might think that's a bit lame, but then have you ever noticed a giant unrecognisable creature floating nearby with the slight hint that its main source of food is lost sailors, and would you venture near it if you had? And so it was that this vast, unknown creature remained just off Scalloway for two weeks before it disappeared from both sight, and recorded history during a violent storm.

Whatever this mystery sea lump was we will never know. It could be that it was the remains of some ship that had foundered. After all what could look more like an upturned keel than, say, an upturned keel? But, if it was an animal, what likely candidates do we have? While it could have been a whale, it's hard to see why no one would have recognised it as one. Perhaps then it could have been the remains of some giant squid, but is that possible? It is difficult to say for certain, but Shetland has played home to some very large squid strandings.

Sometime around 1860 or 1861 a 16 foot squid washed up between Hillswick and Scalloway. Maybe not the biggest giant squid to come ashore, but a fair size you have to admit. But that isn't the biggest squid to come ashore in Shetland, that honour goes to a 24 foot long squid, *Architeuthis monachus*, that was found at the head of the Whale Firth on the island of Yell on 2nd October 1949. This is actually the largest recorded squid so far to be found in British waters, and certainly that's big, but could something like that - only bigger - really have turned up in 1810? It's unlikely, but certainly not impossible. In 1878, across the Atlantic in Newfoundland at a place known as Thimble Tickle Bay, a squid was washed up that was said to have been over 55 feet in length. Now that's getting bigger, and there are some people who consider that there are far bigger squid out there, so perhaps it was one of these that floated outside Scalloway all those years ago. Whatever the truth of it we will most likely never know for certain. If only someone had braved rowing out to have a closer look.

Stornoway 1821

Sometimes we only get a glimpse of a mystery sea serpent or lake monster, and so it was in this case.

All the details regarding this creature come from a letter published in the *Inverness Courier* concerning a visit the writer of the letter had to the isle of Lewis in 1821.

At the time there was a report of a large lake monster swimming around the lochs area. While the gentleman may have failed to shoot this particular creature, he did hear about another one that had been caught.

> *Inverness Courier*
> 'Sir,
> I observe a statement copied from your journal relative to a monster which has been seen in one of the lochs in Lews [sic] Island.
>
> 'I beg to inform you that when shooting in that island in September 1821, with four gentlemen, we saw the same animal and probably in the same loch; and for several hours endeavoured to get an opportunity of shooting at the creature, but without success. We dined that evening with Mr. Mackenzie, at Stornoway, and mentioned what we

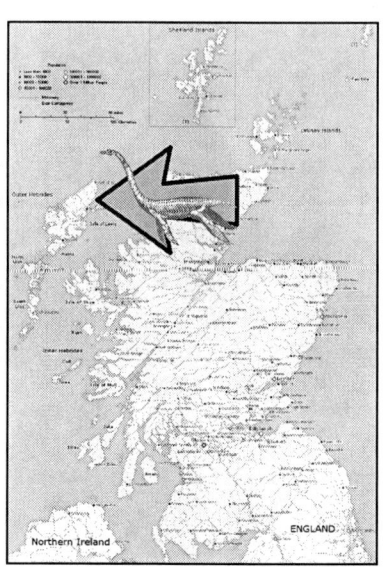

had seen to him. Mr. Mackenzie expressed considerable surprise, but stated that the report was current in Lews [sic] Island when he first came there, that such an animal had been captured in that very lake, and that it resembled in appearance a huge Conger eel and it required one of the farm carts to convey it to Stornoway. The capture of this creature must have occurred seventy or eighty years ago.
I am, Sir your obedient servant, W.P.'

Unfortunately that's all we know about this large mystery eel, but that sadly is often the way with early sightings.

Benbecula 1830

Of all the reports of deceased sea serpent sightings contained in this book, this is the only one where the demise of the creature is directly attributed to human intervention. If we are to take the evidence at face value it also asks the shocking question about how close the creature killed was to being human. For unlike many a Bigfoot hunter, the folk in this tale had no qualms about killing the creature. And what is this creature? Well it is nothing less than a mermaid. Having already had a look at a report of a giant woman washed ashore in the dim and distant past, let us now look at a much smaller fish-tailed one who washed ashore on the Outer Hebridean island of Benbecula in 1830.

Benbecula certainly has had more than its fair share of strange things over the years from its globsters and faery dogs to the sneaky Hebridean fox, but perhaps its most celebrated tale is that of the mermaid. Well possibly that claim could just as likely be made for the tale of the flight of Bonnie Prince Charlie in 1746, and while that tale may have plenty of adventure and cross-dressing it isn't for this book.

Well our tale takes place eighty-four years after Charles's visit in or about the summer of 1830 off the coast near Sgeir na Duchadh near Grimnis. At the end of a long day of backbreaking seaweed cutting, a group of locals were getting ready to return from the beach to their black houses. One of the women in the group headed down to the lower end of the reef, from where they had been collecting the seaweed, to wash her feet. As she finished and was about to put on her stockings, she was surprised to hear a splash in the calm sea, and, intrigued, she looked around to see what had caused the disturbance. Upon raising her head she was somewhat shocked to see a creature that she could only describe as looking like a miniature woman swimming just a few feet away. Understandably this wasn't something that you saw every day, not even in the rather mystery-haunted land that was nineteenth century Benbecula.

Her stockings now completely forgotten about, the woman called out to the rest of the group '*maighdean nan tonn*', a name translating as 'maiden of the waves', in other words a mermaid. Whatever she said it didn't take long for the rest of the seaweed cutters to be drawn to the spectacle of the little lady playing in the water, seemingly oblivious to the attention being paid to her, but then again when have mermaids ever really been the shy, retiring types? Even this demure one was observed to be performing somersaults so hardly failing to draw attention to herself. Eventually she settled down calmly to comb her hair in front of the gathered group, well if you're a temptress of the deep it's probably best to look after your hair because you never know who might be watching.

But like the mermaids of the old seafarers' tales this one was a temptation, no doubt all that hair combing, and not forgetting the topless cavorting, goes with it. Whatever the attraction, a number of the men in the group decided that they would attempt to get a little closer and try if possible to catch the creature. But even as they waded into the cold water out towards her she would drift just a little further from reach still playing around, still enticing the men further out. Perhaps that old mermaid gene was kicking in, and regardless of the depth of the water the men continued to wade in after her. Back on the beach, the crowd was starting to grow a little larger as the village children joined in to see the commotion.

Now while the men of the group may have been prepared to wade and swim out after the mermaid, and the women folk might have been prepared to just stand by and watch them drown, the local children had other ideas. It seems that the good folk of Benbecula had raised a batch of Neds (Non-education Delinquents, a rather delightful Scottish term for youthful wastrels with a little too much time on their hands) for they weren't fooled by the mermaid's glamour. They just followed their natural instincts, no doubt fortified by the odd bottle of Buckie (that's Buckfast Tonic Wine to the uninitiated, the tipple of choice for those who require cheap, strong alcohol), and they just picked up some of the large pebbles that littered the beach and started to throw them at the cavorting mermaid.

Perhaps because she was spending all her time teasing the men, she wasn't aware of the new threat that was flying through the sky towards her. The majority of the stones splashed harmlessly in the sea around her, some overshooting, some falling well short, but one of the stones must have had Goldilocks' name on because it was just right - if not deserved. That one stone struck the mermaid in the back and, as if pole-axed, she slipped below the waves, no doubt to the pursuing men's annoyance, and if you think about it the mermaid probably wasn't all that impressed either. It was probably a good job for the Neds that it was a west coast merbeing, because if it had been one of the east coast blue men they would have probably been throwing the stones back, much rougher creatures them but that's another tale.

Their fun over, the group went back to their seaweed cutting no doubt discussing what they had been privileged to have seen, well maybe the children went off to sup some more Buckie and hang around awaiting the introduction of the bus shelter.

If that wasn't mystery enough, the tale takes on a darker tone when a few days later a body was washed ashore a couple of miles from where the mermaid had been sighted happily playing. So it was that the body ended up on the beach at Cuile, near the township of Nunton.

Now if you are an old romantic you might like to think that the body washed up had nothing to do with the mermaid that had so happily been playing in the sea near the seaweed cutters. After all, plenty of things wash up on the shore as it is without the need to add to the list a recently and recklessly killed mermaid. Perhaps the stone only annoyed the mermaid and she had just gone off in a huff. If that's what you prefer fair enough, a happy ending to the story.

If however you don't believe in the coincidence of a mermaid being sunk by a stone and a completely different woman with a tail being washed up on nearby beach, the rest of the tale might be of interest.

So it was a couple of days later a bedraggled body was washed up on the shore with the tide. During the ensuing day after the first sighting, word had certainly gotten around the area and there was no doubt in anyone's mind that this was the poor unfortunate mermaid who only a few days before had been seen splashing around until her cavorting was stopped by a well-aimed rock in what was certainly a terminal way.

Word got round and reached the keen ears of the sheriff of the district, the baron-bailie and the factor for Clanranald Mr Duncan Shaw. By the time Duncan Shaw turned up at the scene a large crowd had gathered around the prone body of the mermaid, no doubt attracted by stories of the earlier antics of the Babe of Benbecula.

Recording these events seventy years later Alexander Carmichael in his book *Carmina Gadelica* described what the crowd found lying dead at their feet.

> The upper portion of the creature was about the size of a well-fed child of three or four years of age, with an abnormally developed breast. The hair was long, dark and glossy, while the skin was white, soft and tender. The lower part of the body was like a salmon, but without the scales. Crowds of people, some from long distances, came to see this strange animal, and all were unanimous in the opinion that they had gazed on the mermaid at last.

When Sheriff Duncan Shaw arrived at the scene he ordered that the mermaid be given a proper burial, after all, it was the least that the community could do after having killed her. So it was that the mermaid ended up being dressed in a shroud and placed in a small coffin that had been made for her. Once all this had been done the body was moved a little further inshore from where it was found and buried; there was even a good turn out for the impromptu funeral - whether they were there as genuine mourners, out of shame, or just morbid curiosity we may never know.

But the story doesn't end there because ever since that date people have been trying to find this body, possibly the best proof that mermaids are real. That is it would be the best evidence if anyone could find the remains. When Carmichael was writing in 1900, he stated that there were still people alive who had been lucky enough to both see and even touch the mermaid. But as time went on those people eventually died and the location of the burial site of the mermaid started to be forgotten. Even as late as the 1960s visitors could still be pointed to the grave of the mermaid in the graveyard at Nunton.

It was in this burial ground, the graveyard of the chapel of St Mary, that the author R. MacDonald Robertson claims to have seen the location of the grave. As the graveyard is east of Cuile Bay, it would lend itself to the belief that this could be the final resting place especially if the locals gave the mermaid a Christian burial.

But as it is with these things there are plenty of other people with different ideas of the final resting place.

In March 1994, Dr Shelagh Smith from the Royal Museum of Scotland discovered what appeared to be a headstone in the sand dunes at Cuile Bay. This led the archaeologist Adam Welfare, in August of the same year, to study the stone only to decide that it wasn't a headstone after all. It's recorded that a person by the name of MacPhee told that his family had a traditional belief that the lifeless body of the mermaid had been washed ashore at the south of Cuile Bay on a rocky inlet known as Bogha mem Crann (stinky bay), so called as it is a place where fermenting seaweed gathers.

The latest attempt to find this missing vital clue was as recent as 2008 when the American cryptozoologist Nick Sucik spent a few weeks on the island investigating the story. He found a grave-shaped mound in a field that - according to some of the locals - could be the fabled burial site. So far no digging has taken place, but who knows? It might not be long before we have a firm answer to what washed up on that beach all those years ago.

But in the continued absence of the body, let's speculate on what it could have been. It is worth considering that the people living in the area would be well versed in the sea creatures that were to be seen locally and they would certainly be well able to spot the difference between the washed up remains and those of, say, a seal or an otter, so could it be that it really was a mermaid?

Of course there is one other possibility that has struck me since I first heard the story: what if the body they found wasn't the same creature as the mermaid that had been seen playing around in the sea by the seaweed cutters, but a fake.

Now it is not impossible that at the time there was someone living on Benbecula who may well have sailed a long way from the island before returning, and may well have had the opportunity to purchase a Jenny Haniver, a composite mermaid produced for the man who has everything. So hearing the tale he takes a trip down to the beach one morning and leaves the 'mermaid' where it can easily be found, with the tales of the recently sighted mermaid still fresh, two and two are added together and hey presto!

But that would just spoil the story wouldn't it and I, for one, look forward to the day when that grave is finally found and the answer is finally made. And I, for one, hope the Babe of Benbecula is a real *maighdean nan tonn*.

Firth of Forth 1848

Our next encounter with a supposed sea serpent took place in the Firth of Forth. The Firth is the estuary of the River Forth where it flows into the North Sea, between Fife to the north and Edinburgh and West Lothian to the south. Or, if you are on planning coming into Edinburgh Airport from the isle of Lewis, it's the big bit of water that you fly all too low over.

It was long before the days of cheap flights when, in 1848, a fishing smack going by the name of the *Sovereign*, sailing out of Hull, was to be found fishing in the Forth for Lord Norbury. Along with the usual catch of fish the crew managed to catch something rather more unusual. They managed to capture in their haul a very long, serpentine fish. It was recorded that the fish, when laid out flat, easily exceeded the length of the vessel with the extreme parts of the fish extending over the edges of the craft at both the stem and stern.

Now that does sound very impressive with modern estimations of the length of a fishing smack ranging anywhere from 35 to 60 feet in length. Well you would think it was very impressive, but to the crew of the *Sovereign* it appears that it was nothing special as they had previously, on another fishing trip, come across a bigger example that was reported to be dark in colour. The current fish they said was only 4 to 9 inches thick with a dorsal fin just 7 to 8 inches high. The crew, it seems, easily identified the fish as an oarfish. The crew's damning verdict of the creature was that it was a fish that in Scandinavia not even the dogs would eat. Whether that really tells us anything about the oarfish, or if it tells us more about the fussy eating habits of dogs in Norway and Sweden is a debatable point.

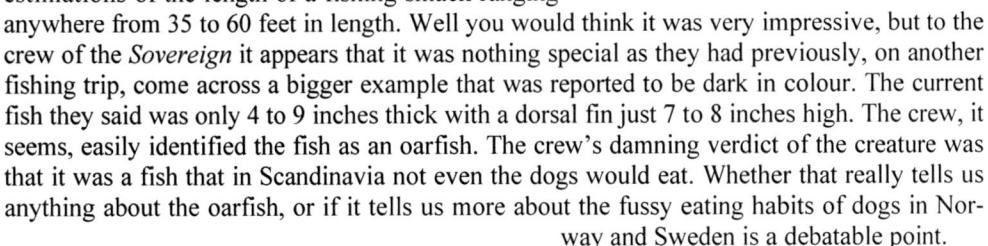

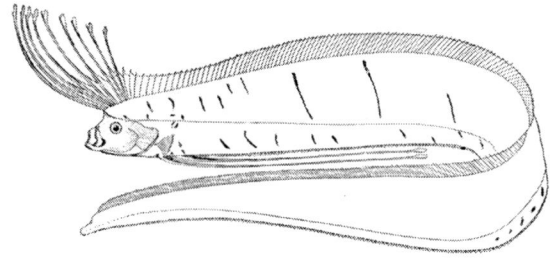

So it appears that this sea serpent carcass wasn't anything but a very interesting fish. The oarfish has been recorded to reach lengths up to 36 feet and is the longest bony fish known, it is, however, rarely seen alive or served up to dogs these days.

Usan 1849

Once again we find our next supposed sea serpent while out with the fishing fleets, this time the boat was from Usan (Angus). It was noted in the *Zoologist* of 1849 that the following had been reported in the Scottish newspaper the *Montrose Standard*. The article was as follows.

> *Montrose Standard*
> A young sea-serpent.—On Friday, while some fishermen belonging to Usan were at the out-sea fishing, they drew up what appeared to them a young sea-serpent, and lost no time in bringing the young monster to the secretary of our Museum. The animal, whatever it may be called, is still alive, and we have just been favoured with a sight of it; but whether it really be a young sea-serpent or not, we shall leave those who are better acquainted with Zoology than we are to determine. Be it what it may, it is a living creature, more than 20 feet in length, less than an inch in circumference, and of a dark brown chocolate colour. When at rest its body is round; but when it is handled it contracts upon itself, and assumes a flattish form. When not disturbed its motions are slow; but when taken out of the water and extended, it contracts like what a long cord of caoutchouc would do, and folds itself up in spiral form, and soon begins to secrete a whitish mucous from the skin, which cements the folds together, as for the purpose of binding the creature into the least possible dimensions.

There was a correspondent to this article, an E. Newman, who thought that the animal was a specimen of *Gordius marinus*.

All well and good, but it appears that the name has been applied to two different marine worms. One, *Gordius marinus montagu*, averages half an inch in length at maturity and lives parasitically in the entrails of some fishes, especially herrings; while the other – now called *Lineus longissimus* – is one of the longest known living animals, with a specimen from 1864 exceeding 155 feet.

Once again it appears that the sea serpent in this instance was anything but a sea serpent, in fact it was a worm. But, which of the two candidates do you think they may have recovered from the deep? I for one wouldn't like to see a massive parasitic worm; it was bad enough that whatever it was, was excreting white mucus.

In case you were wondering, caoutchouc is a name for natural rubber, or India rubber. Now I bet the description of the creature makes a lot more sense now, even if it makes the creature even less appealing.

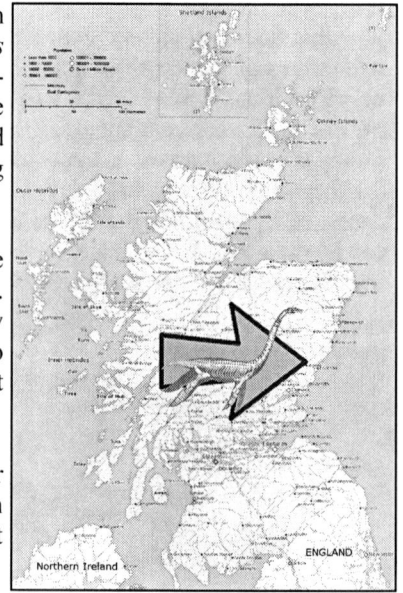

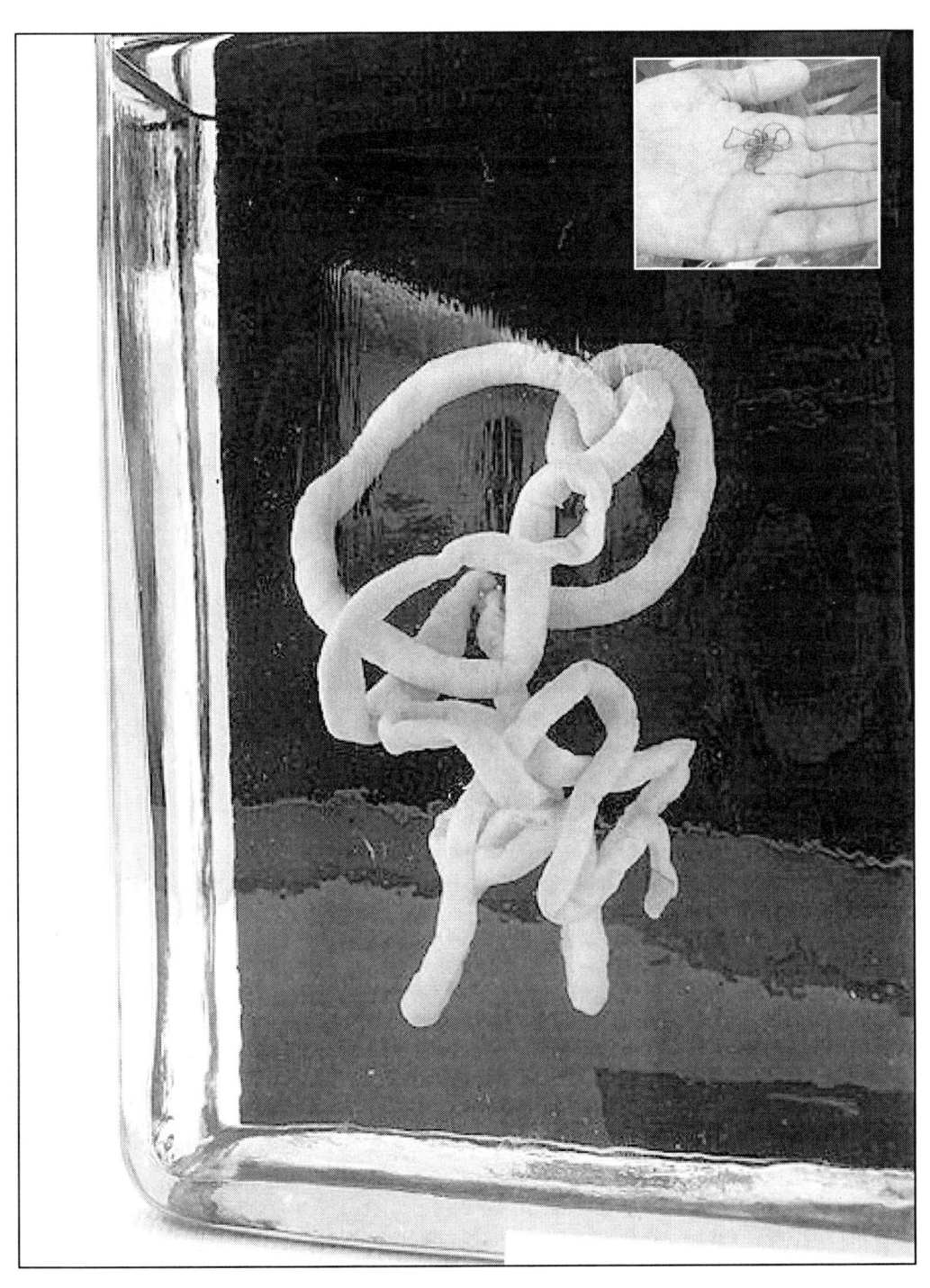

The two creatures that have been known as *Gordius marinus*. The main image is
Lineus longissimus, and the insert *Gordius marinus montagu*

© Citron / CC-BY-SA-3.0

Griais 1851

Occasionally the whole sea serpent doesn't wash up and the remains that are left for us to look at are just a few parts of the mystery animal, which isn't surprising when we consider that the usual occurrence, when a sea serpent was sighted, was to run for your gun and take pot shots at the unfortunate creature. Such is the next case we look at from the hamlet of Griais on the isle of Lewis.

Like many of these tales of mystery animals in the nineteenth century, the remains had no trouble making it into the letters pages of *The Times*. I wonder if it would be as easy to get a letter reporting a sighting of a sea serpent into it nowadays. Whatever the present challenges of such a feat, back in 1893 the Scottish physician W.M. Russell had no problem in getting such a letter published. Then again perhaps he had been trying for some years because he had found out about the sighting in 1851 when he had been in conversation with a Mrs Maciver at her home at Griais.

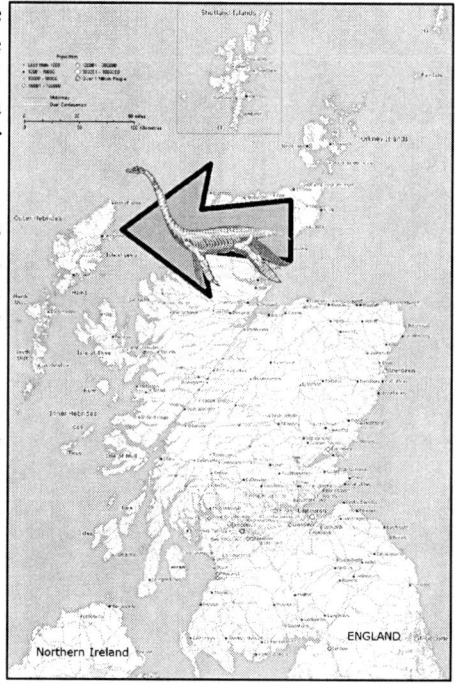

While her sighting had happened some time before Mr Russell's visit, there is no firm date given for Mrs Maciver's sighting other than around 1851, but the details were still fresh in her mind. She described how on the day in question she had been looking out from her house down and across Broad Bay to the peninsular of Point, and while there might normally be nothing more to see than the usual fishing boats, this day was different because there, near to the shore, she could see a great commotion caused by fish leaping out of the water as they tried to escape a great sea serpent that was pursuing them; if nothing else this sighting suggests what a sea serpent's main diet is. As if Highland tradition demanded it, the local men ran for their guns and as soon as they had them loaded proceeded to shoot at the monstrous beast. Mrs Maciver was certain that the men had wounded the unfortunate creature, but I would have to differ with her on that point. Can you imagine for one second that anything short of coastal artillery would even dent such a creature. The much-maligned sea serpent headed for the reef of Sgeir Leathann where it proceeded to raise not only its head, but its body, some eight to ten feet up onto the rocks where it appeared to scratch some itch, perhaps dislodging the odd bullet that might have dented a scale or two. It rested upon the rocks for a short time until it was interrupted by the pursuit of the emboldened gun-toting locals who perhaps hoped to finish the animal off. No doubt the sea serpent, thinking 'bugger that', slid off the rocks and set off for deeper water and away from the pursuing boats, leaving not only a great wake behind it but, as the men would soon discover - on the rock where the sea serpent had been scratching - a number of scales that they took to be from the creature. Some of these scales would make their way to Mrs Maciver who would in turn pass them on to W.M. Russell, who would promptly lose them sometime before 1893. Despite his careless ownership, Russell did at least describe them before they disappeared for good as being both the shape and size of scallop shells. Let's hope that Mrs Maciver wasn't pulling the wool over his eyes with a couple of cheap ashtrays.

Loch Ness 1868

For our next sighting, we head inland to perhaps one of the most famous haunts of a mystery creature, Loch Ness. Nessie is easily Scotland's most famous cryptid with its current rise in popularity starting in 1933. That isn't to say that sightings of a mystery lake monster were not reported from a much earlier date.

It is recorded that St Columba himself in the sixth century had a run-in with the beast[*] and there have been various stories throughout the following centuries.

But for all the many sightings of a strange creature swimming about the loch, or posing for the passing photographer, there is one question that appears never to have been successfully answered: where is the body?

Well the first time that we find a report of a body washing ashore on the loch is way back in the winter of 1868. The *Inverness Courier* of 8th October 1868 reports the following.

> *Inverness Courier*
> 'a few days ago a large fish came ashore on the banks of Loch Ness

---

* According to Wikipedia: "The earliest report of a monster associated with the vicinity of Loch Ness appears in the *Life of St. Columba* by Adomnán, written in the 7th century. According to Adomnán, writing about a century after the events he described, the Irish monk Saint Columba was staying in the land of the Picts with his companions when he came across the locals burying a man by the River Ness. They explained that the man had been swimming the river when he was attacked by a "water beast" that had mauled him and dragged him under. They tried to rescue him in a boat, but were able only to drag up his corpse. Hearing this, Columba stunned the Picts by sending his follower Luigne moccu Min to swim across the river. The beast came after him, but Columba made the sign of the cross and commanded: "Go no further. Do not touch the man. Go back at once." The beast immediately halted as if it had been "pulled back with ropes" and fled in terror, and both Columba's men and the pagan Picts praised God for the miracle.

Believers in the Loch Ness Monster often point to this story, which notably takes place on the River Ness rather than the loch itself, as evidence for the creature's existence as early as the 6th century. However, sceptics question the narrative's reliability, noting that water-beast stories were extremely common in medieval saints' *Lives*; as such, Adomnán's tale is likely a recycling of a common motif attached to a local landmark".

Others have interpreted this story as an account of a face-off between the saint and a tribal shaman.

about two miles to the west of Lochend Inn. Neither the name nor the species of the strange visitor could be satisfactorily explained, and large crowds of country people went to see and examine for themselves, but left without being able to determine whether the monster was aquatic, amphibious, or terrestrial. Some of the most credulous natives averred that a huge fish, similar in size and shape, had been occasionally seen gamboling in the loch for years back, and with equal determination protested that its being cast dead on the shore boded no good to the inhabitants – that, in fact its presence presaged dire calamities either in pestilence or famine, or perhaps both.

At last, an individual better skilled in the science if ichthyology appeared on the scene and ascertained that the strange visitor was nothing more or less than a bottled nosed whale about six feet long.

How one of the denizens of the ocean came to be cast ashore at Loch Ness was the next question, but, this too, has been set at rest, for it was ascertained that the blubber had been taken off!

The fish had, of course, been caught at sea, and had been cast adrift in the waters of Loch Ness by some waggish crew to surprise the primitive inhabitants of Abriachan and the surrounding districts. The ruse was eminently successful.'

So there we have it, the first recorded Loch Ness corpse hoax, and as you will come to find further on in the book, it isn't going to be the last. This time it appears that the crew of a passing boat had managed to get their hands on the corpse of a northern bottlenose whale.

There are a couple of things that strike me about this sorry tale, one is the fact that once again country folk travel to see the remains of a serpent, the other is the implication that there was something rumoured to be swimming about in Loch Ness, after all why fake a monster in a location without an ongoing tradition?

Isle of Man 1872

Occasionally something is recovered from the sea that appears to be a sea serpent at first glance only for it to turn out to be something rather more mundane.

The creature that is most often blamed for this is the basking shark, or rather the decaying remains of one, and sometimes the culprit is a great big lump of rotten whale (if not as the previous report of a skinned whale). However sometimes the sea serpent is nowhere as interesting or as smelly, which is a real bonus for your sinuses.

So here is the story as reported on 1st April 1872 in the *Glasgow Herald*:

Sea Serpent Carcasses—Scotland

Glasgow Herald
The Great Sea Serpent Caught At Last
Many persons are yet sceptical as to the existence of a real sea serpent, and have considered the tales they have heard concerning it either fabulous or the effects of a disordered vision. But as the present sea serpent got land-trapped, it was bound to be caught, and after examination, if it has not turned out to be the veritable one, it is most certainly a fac-simile, in the shape of a log of wood 18 feet long and about 12 inches in diameter, literally covered in barnacles, each from 10 to 18 inches in length. The mass when floating between wind and water, and pitching up and down, would undoubtedly appear a formidable sea monster, as the barnacles look like scales.

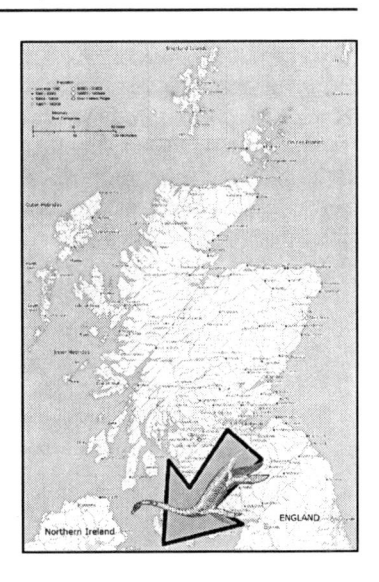

The marine curiosity was picked up by John Kermode and his crew, and has been exhibited by them in a shed on the east Quay, Ramsey.—Mona's Herald.

Okay the sighting wasn't in Scotland, but the Isle of Man is near enough to show the general interest in sea serpents at the time. While the tale highlights a possible source of many a sea serpent sighting, it also shows more importantly that people really will pay good money to go and view a log in a shed.

Oban 1877

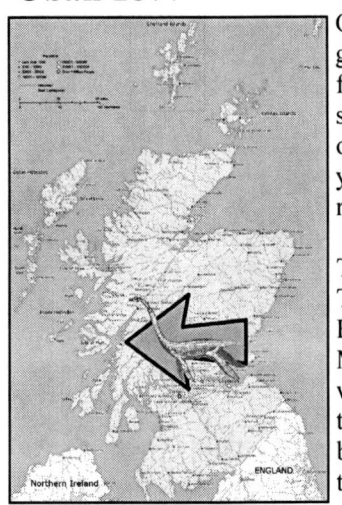

Our next sighting took place in the coastal town of Oban 'the gateway to the isles' so called because that's where you get the ferry from if you head off to explore the Hebrides. If you are setting off into the Hebrides I recommend that you take a copy of *Mystery Animals of the British Isles: The Western Isles* with you, it will give you something to read while you wait for the rain to stop falling.

The town has a couple of claims to fame. The first is McCaig's Tower, a rather impressive folly that looms over the town from Battery Hill. The folly was commissioned by John Stuart McCaig, and erection of the tower started in 1897 and ceased with the death of McCaig in 1902. It is said that the design for the tower is based on that of the Colosseum in Rome and was built to be a lasting monument to the McCaigs. It would be fair to say that if you're in Oban you can't easily miss it, as for a

Sea Serpent Carcasses—Scotland

start its circumference is 656 feet with two tiers of arches. Of course, it was built a few years after our sighting but you can hardly mention Oban without a mention of the folly. The town's latest claim to fame is its failed fireworks display. In 2011 a planned fireworks display for bonfire night went slightly pear-shaped when the £6,000 of fireworks planned to be used in a 30 minute display unfortunately all set off at once, resulting in them all going off within 50 seconds. But, if the next report is to be believed, it's not the only time when there have been plenty of fireworks in Oban. In this report from the *York Herald* it appears that a running gun battle with a sea serpent went on for most of the day and night.

> *York Herald* Tuesday 1st May 1877
> Capture of the sea serpent.
> The *Glasgow News* publishes a circumstantial narrative by a resident at Oban, from which, if it be true, it appears that the sea serpent has at length been actually captured at the place. Under date of 27 ult., the correspondent writes :- 'A most extraordinary event has occurred here which I give in detail, having been witness to the whole affair. I allude to the stranding and capture of the veritable sea-serpent in front of the Caledonian Hotel, George Street, Oban. About four o'clock yesterday an animal or fish, evidently of gigantic size, was seen sporting in the bay near Heather Island. Its appearance evidently perplexed

Oban at time of serpent fight, Hans Gude Oban Bay,Scotland 1889.

Oban Painted Sir James Guthrie 1893.

a large number of spectators assembled on the pier, and several telescopes were directed towards it. A careful look satisfied us that it was of the serpent species, it carrying its head fully 25 feet above the water. A number of boats were soon launched and proceeded to the bay, the crews armed with such weapons as could be got handy. Under the directions of Malcolm Nicholson, our boatman, they headed the monster, and some of the boats were within thirty yards of it when it suddenly sprang half-length out of the water and made for the open. A random from several volunteers with rifles seemed to have no effect on it. Under Mr. Nicholsons's orders, the boats now ranged across the entrance of the bay, and, by the screams and shouts, turned the monster's course, and it, headed directly for the breast-wall of the Great Western Hotel. One boat, containing Mr. Donald Campbell, the Fiscal, had a most narrow escape the animal actually rubbing against it. Mr. Campbell and his brother jumped overboard, and were picked up unhurt by Mr. John D. Hardie, saddler, in his small yacht, the Flying Scud. The animal seemed frightened, and as the boats closed in the volunteers were unable to fire, more, owing to the crowds assembled on the shore. At a little past six the monster took the ground on the beach in front of the Caledonian Hotel, in George-Street, and his proportions were now visible. In his frantic exertions, with his tail sweeping the beach, no one dared approach. The stones were flying in all directions; one seriously injuring a man called Baldy Barrow, and another breaking the window of the Commercial Bank. A party of volunteers, under Lieutenant David Menzies, now assembled, and fired volley after volley into the neck according to the directions of Dr. Campbell, who did not wish, for scientific reasons, that the head should be damaged. As there was bright moon this continued until nearly ten o'clock, when Mr. Stevens, of the Commercial Bank, waded in and fixed a strong rope to the animal's head, and by the exertions of some 70 folk it was securely dragged above high-water mark.

Its exact appearance as it lies on the beach is as follows:- "The extreme length is 101 feet, and the thickest part is about 25 feet from the head, which is 11 feet in circumference. At this part is fixed a pair of fins, which are 4 feet long by nearly 7 feet across at the sides. Further back is a long dorsal fin, extending for at least 12 or 13 feet, and five feet high at the front tapering to 1 foot. The tail is mere of flattened termination to the body proper than anything else. The eyes are very small in proportion, and elongated, and gills of the length of 2½ feet behind. There are no external ears; and as Dr Campbell did not wish animal handled till he communicated with some eminent scientific gentlemen we could not ascertain if there were teeth or not. Great excitement is created, and the country people are flocking into view it. This morning Mr Duncan Clerk, writer, formally took possession of the monster, in the rights of Mr McFee, of Appin, and Mr. James Nicol, writer, in the name of the Crown.'

Well what do we make of that? Once again we note that country people flocked to it, and that is never a good sign considering the last time they gathered they were on their way to look at a skinned whale on the shores of Loch Ness. Did I mention Oban's other claim to fame is its distillery? I am not saying that it has any bearing on this story, but when we have a look at the follow-up report you have to wonder.

> *Jacksons Oxford Journal* Saturday 5th May 1877
> Mr. Robertson, manager of the Royal Aquarium, states that as soon as he had read the above statement he telegraphed to Mr. Duncan Clark, at Oban, offering to purchase the serpent for exhibition and received the following reply :- 'The whole thing is a shameful hoax, deserving no attention except to punish the author.'

And that is the end of the Oban sea serpent carcass, nothing more than a shameful hoax. Still it's worth repeating if only to get an idea of what a Godzilla film would have been like in Victorian Scotland.

Orkney 1894

It was a few years before another strange creature washed ashore on one of the Orkney isles. Unlike the Stronsa Monster, this rotten lump would fail to really grab the attention of many, that is unless you were down wind of it.

It was September of 1894 when reports of the creature first started to appear in the press, so let's have a look at a couple of them. First stop the report in the *Gloucester Citizen*.

> *Gloucester Citizen* 18th September 1894
> The Miscellany
> In default of the sea serpent, this will do. A Kirkwall correspondent says some sea monster has been washed ashore at Traittland Ronsay. It resembles a walrus, is from ten to twelve feet long, and in a decomposed state. The back of the neck is like that of an elephant; the snout resembles a pig and on the body is grayish hair. Another correspondent says it is thicker than a horse, but is probably much wasted by decomposition. The colour is a light grey. No one seems to be able to say what the species belongs

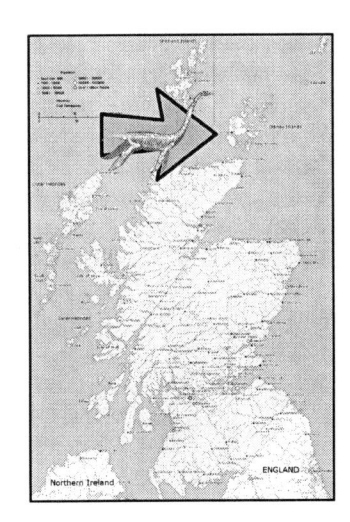

Next stop the *Huddersfield Chronicle*.

> *Huddersfield Chronicle* 19th September 1894
> Death of The Sea Serpent

Sea Serpent Carcasses—Scotland

> The sea serpent is dead. At least it is reported from Kirkwall that a sea monster has been washed ashore at Traittland Ronsay, and is now buried somewhere in the neighbourhood of Kirkwall. A reporter who was present as the interment states that it was about 12ft long, had a back like an elephant, two rows of terrible looking teeth in the upper jaw, a snout like a rhinoceros, and was covered with grey coloured bristles. Even in its death it created great consternation among the natives. Another correspondent says :- It was thicker than a horse, but was probably much wasted by decomposition. The colour was a light grey. No one seems to be able to say what species it belonged.

Interesting stuff you have to admit, but what exactly does it tell us? For a start we notice that the creature had certainly seen better days, so much so that it was probably the advanced state of decay that made identification difficult. Bernard Heuvelmans in *In the wake of the Sea-serpent* identifies the beast as a basking shark, however that may not be the case in this instance. For a start the description mentions teeth, which would rule out a basking shark as a candidate, and with the other descriptive elements in the report we could be looking at the last rotting remains of a vagrant pinniped.

Whatever the identification, the one thing that does stand out are the terrible teeth. Whenever I see the description 'terrible teeth' I wonder if it means that the animal has a mouthful of razor-sharp fangs or just had a mouthful of teeth in dire need of a dentist. I am sure, however, that whatever the state of the teeth they were no way as terrible as the smell.

Caledonian Canal 1899

When the authorities recently drained a stretch of canal in Manchester they found a very strange selection of items in the silt and mud, ranging from the now traditional shopping trolleys and bicycles to the less common sofa, hanging baskets and, perhaps more surprisingly, a chainsaw.* There are, however, even stranger things said to have been found in a drained stretch of another canal, the Caledonian Canal; certainly when they drained it in 1899 they found something very odd.

According to the witness, a Mrs Cameron of Corpach, when the Corpach Canal locks - which are located at the southern end of the Caledonian Canal - were being drained in around 1899, the receding waters revealed a creature resembling an eel, but it was much larger than any eel that they had ever seen and this animal had a

---
* EDITOR'S NOTE: Having worked at the Manchester International II during the glory days of Madchester in 1989/90 I am not surprised at all.

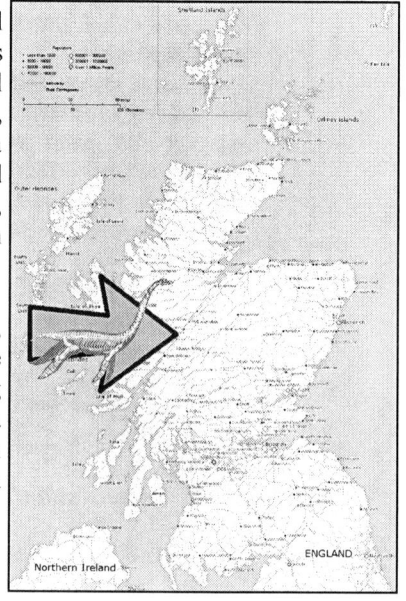

long mane. With that description to tease us there is actually little else to add I'm afraid to say. The question of where the mystery eel had come from has been asked over the years, with some deciding that it was in some way connected to Loch Ness, which lies 30-odd miles up the Caledonian Canal. It is perhaps more likely that the mystery eel arrived via the sea which is just the other side of the Corpach sea lock.

North Atlantic Ocean 1908

While some of the best sea serpent carcasses have been found recently washed ashore, occasionally their remains manage to find themselves in the nets of a fishing boat, and such was the case in the next report.

The following piece appeared in Charles Fort's third book *Lo!*

'In looking over the *London Daily News*, I came upon an item. Trawlers of the steamship *Balmedie* had brought to Grimsby the skull of an unknown monster, dredged up in the Atlantic, north of Scotland (Daily News, June 26, 1908).

The size of the skull indicated an animal the size of an elephant, and it was in "a wonderful state of preservation." It was unlike the skull of any cetacean, having eye sockets a foot across. From the jaws hung a leathery tongue, three feet long. I found, in the Grimsby Telegraph, June 29th, a reproduction of a photograph of this skull, with the long tongue hanging from the beak-like jaws.

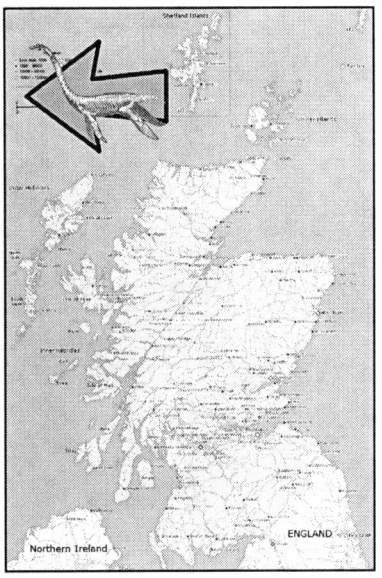

I made a sketch of the skull, as pictured, and sent it with a description to the British Museum (Natural History). I received an answer from Mr. W.P. Pycraft, who wrote that he had never seen any animal with such a skull—"and I have seen a good many!" It is just possible that nobody else has ever seen anything much resembling a sketch that I'd make of anything, but that has nothing to do with descriptions of the tongue. According to Mr. Pycraft no known cetacean has such a tongue.'

And who was this Mr W.P. Pycraft who was corresponding with Fort? Well it turns out he was a curator at the Natural History Museum in London from 1907, and as an osteologist you would think he would know his bones. However, one look at the photo in the paper reveals that the skull was indeed that of a whale. Why Pycraft was unable to identify this may just be down to the sketch provided by Fort, which - you never know - might not have been the best to work from, after all Fort is not known for being an artist.

While this sighting might have been a washout, that isn't to say that ships trawling the North

THE TELEGRAPH, MONDAY, JUN

TRAWLED UP AT SEA.

WHAT IS IT?

The above is a photograph of the strange catch trawled up by the Aberdeen vessel Balmedie (sailing out of Grimsby), whilst on the fishing grounds last week. In the opinion of many, it is the skull, or part of the skull, of some prehistoric monster. The tongue, which is seen plainly in the picture, was well preserved. This curiosity was purchased by Mr. J. Norton.

Sea gather up just faux sea serpents in their nets. In recent years Dutch beam trawlers have had great success in scooping up all manner of extinct animal parts, for example it has been estimated that in the last 40 years they have landed 15,000 mammoth teeth alone.

Dunnet Sands 1934

It seems for a number of years there are no strandings to report from Scotland, but then in the early 1930s things started to change, probably not because sea serpents suddenly started to throw themselves on the shore in vast numbers, but rather because of something that had happened elsewhere in Scotland, the arrival of Nessie. For it was on 22nd July 1933 that one of the world's most celebrated mystery animals decided to announce its arrival by crossing the road that runs alongside Loch Ness, in front of George Spicer and his wife. It was only a year after that sighting that the famed 'Surgeon's Photograph' surfaced like a toy submarine from the still waters of a bath. While today the photo is widely regarded as a fake, at the time it just caught the people's imagination, so much so that the hunt was back on for more lake and sea monsters. As luck would have it, it didn't take long for the first corpse to wash up with reports appearing in various papers. We shall have a look at the report in the *Nottingham Evening Post*.

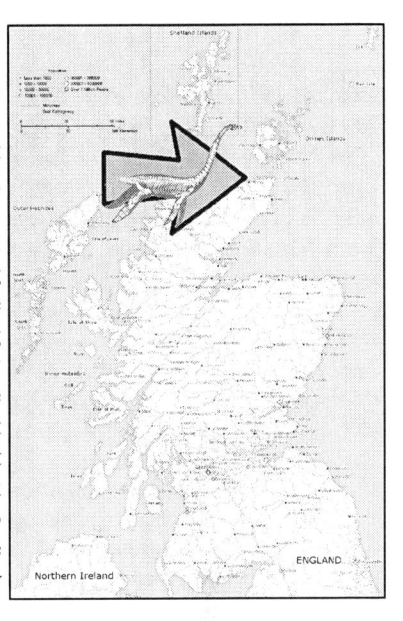

Nottingham Evening Post Monday 16th July 1934
Like Sea Serpent.
Strange monster washed ashore
Mane on the neck
A strange monster resembling a sea serpent has been washed ashore dead on Dunnet Sands, about two miles from Castletown, a village near Thurso, on the northern coast of Scotland.
It has a large flat head, with large eyes, and its chief dimensions are:

Length............................ 29ft
Circumference of head....... 27in
Girth of body (thickest)....... 22in
Nape of neck to tip of fang...25in

There is a mane on the neck, and the body tapers towards the tail. The bones are in sections with connecting sinews, and the largest bone is 18in. in circumference.

Loch Ness vigil
The spot where the creature was washed ashore is about half-way between Inverness and the position of Cape Wrath, where the crew of

the Swedish steamer Nordia saw two monsters on June 27th.

A 10-hour vigil on Loch Ness yesterday by many official and unofficial watchers was without result. Four motorists reported that they saw two monsters on Saturday from Glen Urquhart side one lying still on the surface and the other moving submerging.

Perhaps 1930s journalism at its best, the seamless way it managed to tie the remains of this lump of rotting flesh to both a sea serpent sighting at Cape Wrath in the north, and the Loch Ness Monster. For those not familiar with Scottish geography, the locations are separated by over 100 miles with Cape Wrath located at the top left, Thurso at top right, and Loch Ness way below. It would be fair to assume that whatever monsters were seen by the motorists at Loch Ness at the weekend, they were unlikely to have swum out of the loch and headed up the coast to Dunnet Sands before one of them threw itself onto the beach and promptly started to rot away at high speed, all in little over 24 hours. But what of the Cape Wrath sighting? Could that have been connected to the remains? Well it's worth having at look at how the *Nottingham Evening Post* reported it.

Nottingham Evening Post Thursday 5th July 1934
Another Monster
Giant sea serpent of Cape Wrath.
First taken for wreckage.
A sea monster, 35 ft long and over six feet broad, is reported to have been seen 20 miles from Cape Wrath by the crew of the Swedish steamer Nordia, during a voyage from Finland to Liverpool.
A letter from B.O. Berggren, steward of the Nordia dated 27th, addressed to Stockholm newspaper 'Svenska Daghladet,' reads 'today at 2.35 we sighted ahead of us a formless mass which we first took to be a piece of wreckage, but as the ship slowed down we made out a big sea monster like a giant sea serpent.

"the beast broadened out at the middle. It had a long narrow tail, four great fins, and a something like that of a bullhead – a British fish known also as 'the miller's thumb' from its broad flattened head.

" it was not easy to estimate its size but I reckon it was about 10 to 11 meters long and at least two meters broad.

The first officer who rushed forward watched it glide alongside the ship and then when it was amidships, it turned off and plunging into the sea and disappeared.
"eight persons saw the monster, including a passenger, the principal of a Malmo school.

"we are now lying 20 miles south-west by west of Cape Wrath Scotland"

A rather interesting sighting, and while it compares the sea monster to that of a bullhead, one interesting fact that needs to be considered is that the bullhead, or miller's thumb, reaches a maximum of just 18 cm which leaves it well short of the animal in this sighting.

But is the creature in this sighting in anyway related to the beast on the beach? It has to be said that it is highly unlikely. For a start the dimensions of the two creatures are quite different, and while the body length may be comparable, the width differs considerably, even allowing for a week or so of rotting and being nibbled at by passing fish.

Rather, the connecting of the two creatures in the report highlights something that you often come across in old newspaper reports, and that is that the sea serpent is treated as a singular phenomenon. It would appear to be inconceivable that there could possibly be more than one such creature lurking in the depths, whether that be in the Atlantic, the North Sea or even Loch Ness. The flaw in this approach is that reports then have the danger of being twisted to fit the ongoing narrative.

Moray Firth 1934

For the Loch Ness Monster, the 1930s were a very busy time. If it wasn't putting in personal appearances in the Loch Ness region, so helping to launch a tourist trade that is still strong today, it was lending its name to any odd creature that happened to wash up on a beach, or that got caught in a net anywhere in the world. For this report it didn't have far to travel, just down the River Ness into the Moray Firth, as told in the following article.

Singapore Free Press and Mercantile Advertiser 13th June 1934
Is it the Loch Ness Monster? New Mystery Creature Caught

Sea Serpent Carcasses—Scotland

Lying in a salmon-sorting shed at Findhorn, Morayshire, only 35 miles from Loch Ness, is a strange sea creature which, it is thought, may help to solve the mystery of the loch monster.

It was caught in the nets of Mr. James White, a local fisherman and it is thought to correspond in many ways with the description given by witnesses of the loch monster.

Mr. William Cooper, superintendent of the Moray Firth Salmon Fisheries, who has the strange creature now in his ice shed, said to a reporter:

'it is 13ft long, 14in deep, and about 12in broad. It is of a silvery grey colour and the same width almost the whole length of its body.

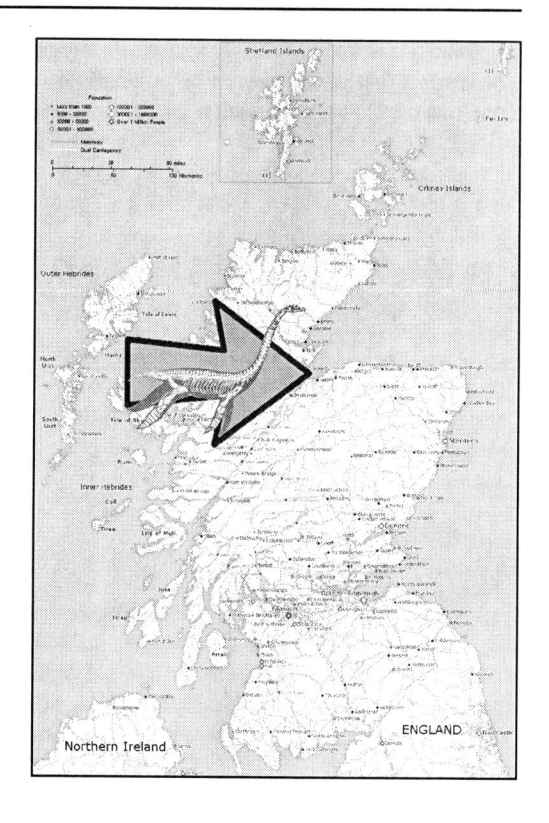

Its head is thin and sharply-pointed. And its eyes are very big.'

Mr Cooper said that it was just possible that it had found its way from Loch Ness into the Moray Firth, on the shore of which Findhorn stands.

'The picture of the loch monster taken by a London surgeon showed what looked like a large fin sticking out of the water.

'such a picture could be taken, I think, if the monster we have found was diving and its tail was just about to go below the surface.'

When a reporter showed a photograph of the Findhorn creature to Professor Graham Kerr, Professor of Zoology at Glasgow University, he said:

'Although I cannot say definitely, it looks very like an oarfish or regalecus.

'to anyone but an expert, a large oarfish swimming in the water would give the impression that it was a sea serpent.'

While the above article does not have a picture of the Moray beast, I managed to track one down in the *West Australian* newspaper, which does indeed show a rather stiff looking fish.

Well there you go then, it was nothing more than a Scandinavian hound's least favourite fishy treat, the oarfish. It is sad to report that the surgeon's photograph of the Loch Ness Monster referred to in the article is now widely believed to be a hoax.

Prestwick 1939-45

As you could well imagine when you are looking into historical records of sea serpent strandings, occasionally some of the details have become lost. We have been lucky that some of the locations for strandings, for instance that of the Stronsa Monster, are easy to find today, while the exact locations where other remains were washed up are a little harder to find. In the latter case, it's usually because of the passage of time. One of the things that makes this next stranding so mysterious is that it happened within living memory, but both the exact location and the date the remains washed ashore have not been recalled exactly.

What we do know about this stranding comes from a letter sent to the American cryptozoologist Dale Drinnon dated 6th June 1977 in which the witness, William Whammond, recalled a sighting that he had had of a strange carcass that washed up over 30 years before.

Whammond recalled how during the Second World War he had been a student studying in Ayrshire when he heard about the stranding of a sea serpent on a sandy beach near to the location of Prestwick Airport, and so he went off to see for himself.

It was following a storm that amongst the other flotsam and jetsam washed onto the beach there was also something a lot more interesting: a sea serpent. Unlike many a supposed sea serpent that has been reported to have washed ashore, this creature's remains were not in an advanced state of decay with flesh hanging off the bone, but were actually in a pretty good condition all things considered (such as the attention of hungry little fishes nibbling and the beaks of ravenous gulls).

What faced Mr Whammond on that beach was indeed something impressive. He estimated the creature to be between 12 and 14 foot in length with the remains consisting of a head, neck, torso and tail. The body

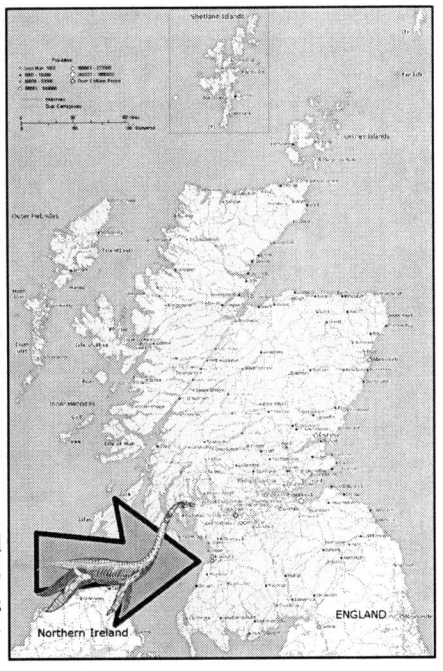

Sea Serpent Carcasses—Scotland

With no known photograph of the Prestwick monster we turn to a 3D model to show this creature in all its glory. You can gauge the size of the beast by comparing it to the intrepid investigator standing near by.

lay flat on its belly with its long neck stretched out in front and a long tail behind it, both extended to a full length.

The head was around a foot long, the neck, torso and tail were roughly equal in length at either

4 or 5 foot apiece. He recalled that the flesh was still soft, and showed no sign of being mummified, rather it was completely fresh with no smell of decomposition to it. In fact the creature was that fresh that if you kicked it, it was like kicking the belly of a fat man (I don't recommend trying that as it's not a nice thing to do). There was no sign of any part of the creature's skeleton exposed; the body was covered in a complete layer of muscle, skin and, interestingly, a short fur similar to that on a seal's skin.

Perhaps we should have a closer look at the remains as they lay there, starting with the head.

Mr Whammond would recall this as being similar in size to that of a horse's head, but that it was broader and flatter with big eye sockets. The creature's eye had by the time of the viewing already disappeared, presumably eaten by scavengers. Other than that damage, the head was still pretty much intact with a working upper and lower jaw. Whammond stated that he opened the mouth and saw that it had teeth but unfortunately, due to the passage of 30 years between the actual sighting and the letter recalling it, he could not remember what they actually looked like. The neck seemed to have been washed about by the tide as it appeared to have left a track in the sand resembling that of a fan.

Turning to the torso, we find the body buried halfway into the sand with the result that while he could see both the pectoral and pelvic girdles and 'joints' showing where four flippers came out of the body, the flippers themselves lay buried in the sand. There was no sign of a dorsal fin on the back, rather a mane that ran down the middle.

Moving on to the tail it was, for want of a better description, like that of a tadpole's tail, that is it had a fin that went all the way around it. This fin seemed to be breaking down into fibres like the bristles on a broom.

And that's what the creature is supposed to have looked like. Whammond recorded that the locals and fishermen who looked at the remains could not think of a creature that it looked like, but he himself thought that the remains looked like they could have been those of a prehistoric reptile, albeit a furry one.

Intriguingly so far, from the description, there is nothing to suggest that the creature was in a serious state of decay, and with its mouthful of teeth it doesn't sound like the usual rotting remains left when a dead basking shark washes ashore. So what could it have been? Sadly, Mr Whammond recalled that the body disappeared. He thought that perhaps it was eventually covered entirely with sand deposited on it during the various tides. Perhaps then some part of this mystery animal still awaits discovery on some Ayrshire beach.

Deepdale 1941

It appears that Orkney is somewhat of a magnet for sea serpent carcasses. We have already looked at the most famous, or at least the most well-known mystery carcass, the remains of the Stronsa Monster that washed ashore during a storm in October of 1808. We have also looked at what could very well have been a very unlucky walrus that came ashore near Kirkwall.

Sea Serpent Carcasses—Scotland

While we had to wait over 80 years between those two creatures being washed up it was less than 50 years when the next creature drifted ashore and much like waiting for a bus, the next one turned up shortly after. So let us have a look at that first one of the new wave, the Deepdale carcass.

To many people if you mention the Deepdale monster they will either look at you blankly or start to talk to you about Preston North End's fearsome new defender. Of course some of the more well read might instantly come up with the correct Deepdale, that is Deepdale Holm on the Orkney mainland overlooking Scapa Flow. It would be safe to say that I used to be in the scary 'new defender' category, but in the course of my research I had heard of the Deepdale carcass as being the twentieth century version of the Stronsa Monster. Whether that description is correct or not you will shortly be able to judge for yourselves.

Unlike the many previous sightings around Orkney, this one took place during wartime, in December 1941, and that may or may not have been part of the cause of the sighting. Since 1914 the Royal Navy had been using the vast natural anchorage of Scapa Flow as a base of operations far enough north to intercept any German fleet that was likely to appear from the Baltic Sea. Of course one of the drawbacks of this is that Scapa Flow might be a great place to anchor your ships but it's a poor place to actually keep them safe from attack. With the threat of enemy submarines ever present, it was decided to hinder the access to Scapa Flow by sinking old freighters in the waters of Holm Sound and Weddell Sound; it was expected that these blockships would be able to stop any naughty submarine sneaking into Scapa Flow. During the First World War this worked. However, at the start of the Second World War the defences were shown to be unsuitable when the German submarine U-47 managed to slip into Scapa Flow by travelling around the side of the blockships. Before it made its way back out to sea, U-47 had managed to sink the Royal Navy battleship HMS *Royal Oak*, which quickly sank, taking 833 of its crew to a watery grave.

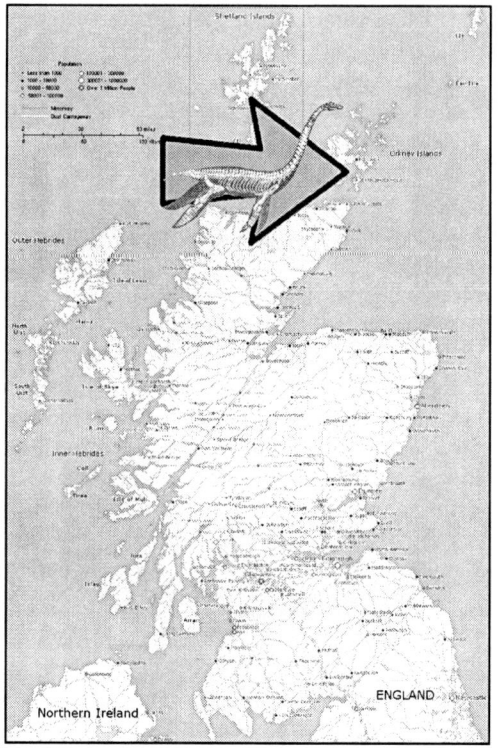

The immediate result of this action would have a lasting effect on the landscape of Orkney; it was decided to replace the blockships with permanent concrete barriers connecting Mainland via the isles of Lamb Holm, Glimps Holm, Burray to South Ronaldsay. This solid barrier would forever cut off the flow of water into Scapa Flow from the east; it also provided a solid foundation for a road connecting the isles, which today makes it a lot easier to see more of them by car than in the past. Work started on these barriers in May 1940, and was completed

After Mr Hutchison

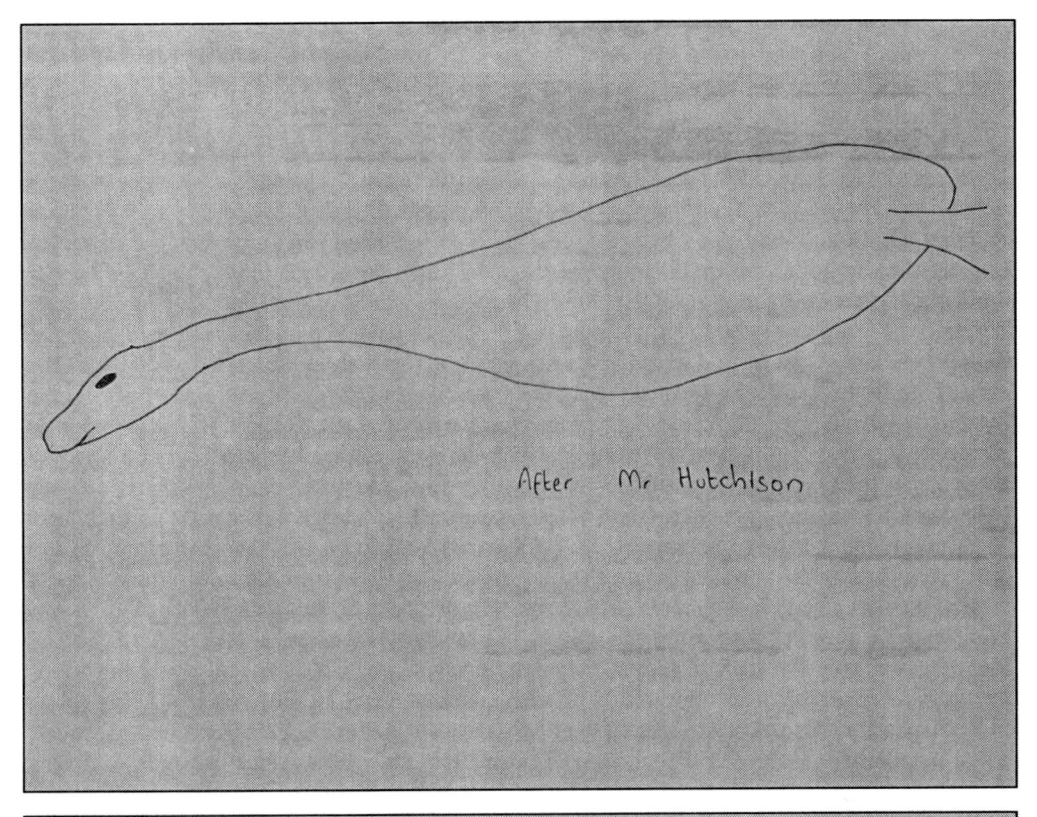

after Provost Marwick

just in time for the end of the war.

It was around Christmas 1941 that the Deepdale monster washed up onto the beach near Deepdale Holm village. At first it appears to have gone unnoticed, at least by the adults. It has been reported that for around three weeks the local children had been playing around the body, but had not considered it worth telling anyone about it. There have been many things said about modern children - getting no exercise, living in a virtual world of computer games - but how desperate do you have to be to have a giant rotting corpse for a playmate, somehow I can't see Disney making a film of that one. Eventually word of the body on the beach made it to the ears of the adults, no doubt intrigued to find out the cause of the story of the big body on the beach, the villagers went down to the beach to see for themselves. You can well imagine their surprise when they were brought face to face with the great rotting lump. When confronted by the monster they were in no doubt about what they saw, and fortunately one person, Police Inspector Cheyne, remembered to bring a tape measure. They measured the remains from head to tail and found it to be an impressive 24 feet 8 inches long, but added that in life it must have been even longer, for it was apparent that part of the head was torn off, and some of its tail appeared to have been broken away. There was hair on the body that was described as resembling coconut fibre in both texture and colour. The monster had a head that was said to be like that of a cow only flatter, it displayed very deep eye sockets, which were three inches in diameter. The creature's neck appeared to be triangular and some 10 foot long and 2 foot round at the thickest point, and at the base of the neck there was a bone shaped like a horse's collar. The animal also possessed four flippers with bone structure similar to that of hands, each was 3 feet 8 inches long and they appeared to have been the creature's primary means of propulsion. As well as being handy with the tape measure, the witness also appears to have been able to estimate weight, stating that there was still at least a ton of stinky flesh on the body.

So far so good, it looks like it's a sea monster. But of course, as is the tradition, it wouldn't take long for someone to come along with a completely different take on what was lying on that beach.

It was around five weeks after the remains of the creature had turned up at Deepdale Holm that the first of the supposed 'experts' turned up and identified the remains as those of a basking shark. Now if you think that would be the end of the story you would have learnt nothing from the case of the Stronsa Monster, which is still going strong a couple of hundred years later.

Now as it happens, there was at that time a man living in Orkney who actually turned out to be one of those rare people, a man who had already seen a sea serpent, in this case that man was Mr W.J. Hutchinson. That would be the same W.J. Hutchinson who had had a close encounter with a sea serpent in the Bay of Meil in 1910, and this sighting, it would appear, made him somewhat of a local expert on the subject. And rightly so, after all not everyone has been lucky enough for an encounter (the full story can be found in *Mystery Animals of the British Isles: The Northern Isles*). So it was that some time after the beast had washed up at Deepdale, Mr Hutchinson was told about it by a friend who urged him to go and have a look for himself. Along with his wife and two sisters, he took the journey to the site. Upon looking at the remains he

Typical basking shark remains

1014

Sea Serpent Carcasses—Scotland

THE ORKNEY BLAST, FRIDAY, FEBRUARY 6, 1942

THE ORKNEY MONSTER — FIRST PICTURES

The carcass weighs between half a ton and a ton and measures over 25 feet. The species of this creature remains a riddle, although some people think that it might be an unusually big basking shark, while others that it is a prehistoric plysiosaurus.

was in no doubt that the creature that lay on the beach attracting seagulls was nothing less than a younger version of the animal that he had seen in the Bay of Meil some thirty years before, and that was most certainly not a basking shark.

Perhaps due to his earlier experience with sea serpents, Mr Hutchinson studied the creature intently and was able to provide some interesting details which, it has to be said, don't entirely match the earlier witness statement.

He estimated that the creature would have been about 20 foot long, with the length of neck from head to body being about 5 feet. The animal was light grey in colour, actually the shade he compared it to was battleship grey. He considered the head to be the same size as that of a Shetland pony and suspiciously similar in shape, but lacking the ears. As I mentioned earlier the remains had been on the beach for a few days at least, and it doesn't take that long for nature to start to dispose of remains. The soft bits are usually the first to go and so it was in this case, as Hutchinson stated that 'the eyes had been picked out by birds and crabs,' now doesn't that sound like it should be in a horror story. However, it was just pointing out the obvious as there were just the bare sockets, roughly one inch in diameter. Now it's worth considering that one difference in the measurements given by the two separate witnesses may well have something to do with the times of their visits. It's possible that the scavenging of animals on the corpse could well have changed the shape noticeably, and we shouldn't ignore the fact that only one of the witnesses carried a tape measure to actually measure the remains.

Back to Hutchinson's description; he mentions that the animal's neck tapered down to the body which he likened in shape to being similar to a seal. He noted that there was no tail, but that on each side there were flippers, which he suggested were the animal's only form of propulsion; he further likened these flippers to those of a frogman. Now in a stunning and practical move he decided to take a sample of the beast for further study, and borrowing a friend's sheath knife he proceeded to cut off a large piece of flipper. Cutting the flesh reminded him of cutting a large skate, but that the flesh was far darker in colour. Sadly, for such an enterprising move, it brought a poor result. Hutchinson took the piece of flipper home and put it in an outhouse, but by the time he went back to it a few days later it had dissolved leaving only a thick skin, which he then threw away.

Along with his friend, a retired chemist, Huchinson considered the nature of the remains; they were both of the opinion that the Deepdale monster was some kind of rarely seen sea creature that they would probably not see the likes of again. So imagine their surprise when it was labelled as nothing more than a rotting basking shark.

Some nineteen odd years after the sighting, Hutchinson was in correspondence with Tim Dinsdale, a dedicated hunter of the Loch Ness Monster, and he was asked if he could add any further details of the sighting. He confirmed that he hadn't examined the mouth for any teeth as his main interest had been in the long neck and head. He did however consider that the mouth was very much like that of a horse. As he couldn't see any nostrils Hutchinson was of the opinion that the animal must have breathed through its mouth.

Another witness to come forward and make a record of her sighting was Margaret Hutchinson. She, along with her sister amongst others, went to have a look at the creature washed up at Deepdale and she also did not agree with the official verdict that the remains were just those of a decaying basking shark. This firm belief was quite reasonably based on the fact that the group had seen sharks washed ashore in many varying states of decomposition and none of them looked even vaguely like the remains on that beach; this was after all the only set of remains to have displayed a neck.

While I am digging up witness reports right, left and centre there is of course room for one more. This one is based on the 'Nature Notes' column in *The Orcadian* newspaper of 29th January 1942 (just across from the table tennis league results) written by a certain Mr J.G. Marwick, former provost of Stromness and, as luck would have it, a competent naturalist. Unlike the earlier mentioned testament regarding the Deepdale monster this one was actually recorded fairly soon after the discovery, certainly around nineteen years before the rest of the reports surfaced.

On the night of 20th January 1942 Marwick was to receive a call from a resident of Graemeshall, Holm who suggest that as he was interested in natural history he might want to take a look at the strange creature washed up at Deepdale. Despite his initial disbelief of the report, he arranged two days later to go with his brother to have a look at the mystery animal. He described the location of the remains as being two and a half miles from St Mary's village. While walking down the track to the beach they soon became aware of the exact location thanks to a huge flock of gulls that flew off at their approach. His first impressions are telling,

> '...and there it lay – well I am at a loss to name it – the remains of some creature, the like of which we had only seen in the imaginative drawings of prehistoric denizens of the deep – call it a monster if you like.'

Like those who were to follow, Marwick made a few observations about the remains on the beach. He noted that the lower jaw was missing and that there was no trace of teeth in the upper jaw. The skull he described as being bare of skin and consisting of gristle, very much resembling that of a skate rather than anything you would associate with the skull of a land animal. He noted that there appeared to be a blowhole in the centre of the skull. As for measurements, Marwick gives a length of 3½ feet of neck connecting the head to the rest of the body. He also seems to be the first person to note the presence of a couple of humps along the back, the first being less than a foot high and the second around 2½ feet high, upon which a few thick hairs could be seen. It seems odd at first that no one else commented upon these humps, but that may have more to do with the state of the body after a few weeks of seagulls and crabs. As for the total length of the body, it measured up to 25 feet from the tip of the skull to where the tail appeared to have broken off. From the view that Marwick had, he was certain it was not the corpse of a whale. As for flippers, he was certain the animal had some, suggesting that the front flipper was over 3 feet long and a foot across, and it was firmly attached to the shoulder of the creature, and that once again it consisted of gristle.

Now he gives a couple of ideas of the beast's weight, he himself suggested that it would be half a ton while his brother opted for it being heavier, somewhere over a ton.

After his first visit, Marwick would return once more to the carcass and would on this occasion talk to Mrs. Anderson whose husband is credited with first finding the remains a week before Christmas. At the time neither Mr nor Mrs Anderson thought that it was anything special, just a dead shark, that is, as Mrs Anderson puts it,

> 'A funny sort o' shark with a neck at both ends'

Yep that would be a very funny shark.

On his second visit, Marwick noted that the carcass had been largely eaten by gulls (you see gulls really do prefer rotting flesh to council tips given the choice). Despite the best efforts of nature to remove the corpse from the beach there was still a great chunk of it left on this second visit, and enough for Marwick to notice one other fact to report, the head had two long whiskers-like things sticking out from the side of it, these antennae he measured to be 5 inches. Deciding that soon there wouldn't be much left, he removed the skull, a vertebrae and the front flipper; all these items he would preserve in a tub, sprinkled in salt.

Following this second visit he had another piece published in *The Orcadian* on 5th February 1942, in which he announces that he considers the remains to be that of a species long considered extinct 'a marine saurian or marine reptile'; his opinion was that it was close to the order of extinct sea creatures the *Ichthyopterygia*. Having made this observation Marwick then goes on to name the creature, and he comes up with the rather catchy title of the '*Scapasaurus*' reflecting the find location of Scapa Flow.

Elsewhere further south beyond the border in England the *Nottingham Evening Post* reported the following.

Sea Serpent Carcasses—Scotland

> *Nottingham Evening Post* Monday 26th January 1942
> The Lighter Side
> A 24-ft 'monster' has been washed ashore on the Orkney Islands. The carcass has not been identified, but it resembles that of the extinct plesiosaurus, a marine animal allied to the lizard and crocodile, and remarkable for the length of neck.
> The skeleton is to be preserved

Whether that report was with regard to the Deepdale remains or our next carcass is hard to say, the dimensions are certainly similar, as you will see when we have a look at the monster from Hunda.

Hunda 1942

Now before I have a look at the evidence for the Deepdale *Scapasaurus*, I think it worth having a look at the other large mystery creature that washed up in the area a short time later, the Beast of Hunda, or the second *Scapasaurus* if you like. It seems that much like buses, you wait for what seems like a lifetime for one and then two turn up.

The Hunda carcass seems to first be reported some fortnight after the Deepdale carcass came to light. It was found washed up on the Hope of Hunda, part of the isle of Hunda, a small uninhabited island in Scapa Flow just off the west side of Burray. At the time of the stranding access to the island had been made easier by the recent completion of a solid barrier boom defence that connects the small island to its much larger neighbour of Burray.

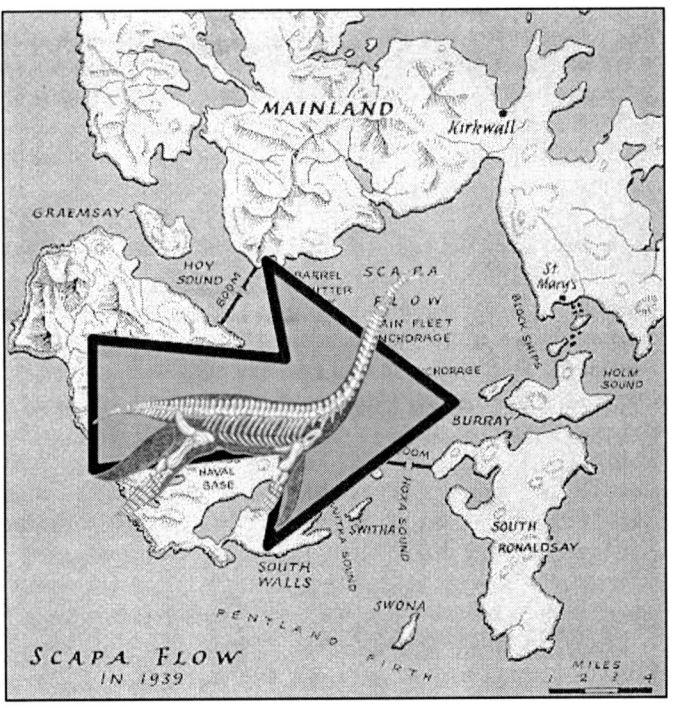

A gentleman by the name of Mr W. Campbell Brodie of St Margaret's Hope went to have a look at the remains; he didn't actually manage to get across to the site until 5th February when he was accompanied by James Macdonald and Andrew Laughton. After making their way across from Burray to Hunda they were all rather surprised to find that the 'monster' really was of the dimensions that rumour had suggested. Campbell would describe the carcass that was lying on the stony beach as a huge elongated yellowish coloured creature wrapped in sea wrack. Once again nature had been doing its best to remove the evidence of the corpse, and by the time that the men arrived the remains consisted of a skull and the vertebral column, and not forgetting the hint of several appendages or fins which appeared to be cartilaginous in origin. There was once again no hint of any lower jaw or signs of teeth. Once again there was a hair-like substance with the texture of coconut fibre that was around 5 inches in length. The creature's skull also featured a large cavity on the raised part that was possibly a blowhole. Now I mustn't forget to mention that this skull, like that of the Deepdale *Scapasaurus*, also possessed two antennae, these measured around 4 inches in length.

Upon measuring the remains, they discovered that it was some 28 feet long, there being 65 vertebrae making up the spinal column. Considering the evidence in front of them the men decided that this was most likely to be another *Scapasaurus* in the same vein as the Deepdale creature. In one of the more random things that has ever happened to exotic unknown animal remains the three men arranged boat hire and had the skeleton moved sections at a time across the water to St Margaret's Hope where it was displayed to raise aid for the Red Cross. You see there are plenty of uses for rotting lumps other than seagull food and fertilizer. Now the *Scapasaurus* might have been doing its thing for charity, but what was the official line? Well according to the Natural History Museum in London, after studying photographs of the remains it showed nothing more that the carcass of a basking shark. However, Marwick wasn't entirely convinced with the official line and he showed the parts of the animal that he had preserved to a lecturer in biology at Durham University. Marwick was pleased to find that the lecturer agreed with him that the remains were of a sea reptile rather than those of a basking shark.

I bet you are starting to wonder what score this game of *Scapasaurus* or shark tennis was going to bring next; well it was actually going to be a large swing in favour of the shark.

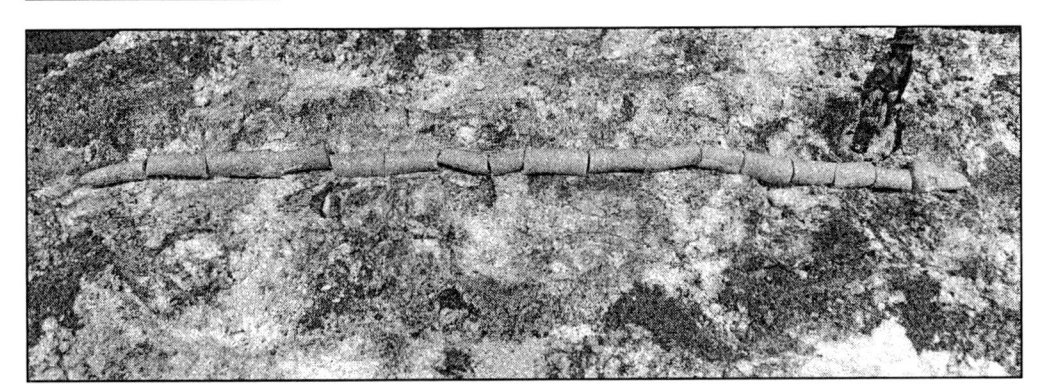

The next people to study the Hunda carcass as it lay in St Margaret's Hope were two members of the Royal Artillery who, thanks to their pre-military qualifications, were seen to be good enough for them to be entitled to give an opinion (what strange times they must have been when artillery men were experts on sea serpents). They noted that the skeleton was entirely cartilaginous and strongly suggested that the remains belonged to a member of the shark family. There was, in their opinion, no true neck, and as for the mouth, or rather the absence of it, they commented that this usually comes adrift from a decaying shark and there is nothing unusual about a missing jaw or gills from such remains. They also mentioned that had the remains been those of a marine reptile then you would have expected to have seen the jaw firmly fixed to the skull. And with that the Hunda *Scapasaurus* was in an instant relegated to being just a very big ex-shark. One down one to go.

Back to the original *Scapasaurus*, the Deepdale monster. Well things hadn't been entirely quiet on that front and Marwick had sent off his collection of preserved parts to the Keeper of Natural History at the Royal Scottish Museum. Unlike many famed remains that get lost in the post, thus forever leaving the identity of a creature unresolved, these salt-preserved bits arrived safely with a Mr Tod at the Museum. On inspection it was announced that the parts were not anything interesting, rather they were the remains of a shark, a big shark, but a shark nevertheless.

And a little aside, there is one comment of the Museum's communication with Marwick that I find highly entertaining,

> 'I'm afraid you must have had rather a hard time with it and endured a good deal of smell.'

Having once worked downwind of a rather large skate that was rotting on the beach outside the office I can fully understand that it wouldn't have been a pleasant experience, but I do suspect that he would have had very clear sinuses before they finally burnt out.

So there you go. It looks like the *Scapasaurus* was well and truly dead and buried, or never existed. Of course in the world of mystery animals all is not always as simple as that and there were still plenty of questions that are asked about the remains. One of the biggest questions asked about the remains was what were the two antennae that both the Hunda and Deepdale creatures possessed? Well it turns out they are not a pair of wobbly sea serpent antennae but are actually rostral cartilages, two rods that support the snout of the basking shark in life. However once a basking shark starts to rot away after death they can take on a new appearance and start to look like little horns. It is not unusual for them to turn up in photographs of dead basking sharks.

There is another question that is asked that is a lot harder to answer and that is why does it appear that not all the witnesses agreed on what they observed on that beach. There was a group who would not accept that the Deepdale monster was a shark, so certain were they that it didn't have any resemblance to any other shark that they had seen.

There is something else about the Deepdale carcass that intrigues me, and that is how the statements given by Inspector Cheyne and W.J. Hutchinson don't seem to be actually talking about the same creature. Certainly the two sets of measurements seem to be well out, and if you compare the illustrations on p.65 which are based on the two sketches of the Deepdale creature you will see that Hutchinson was seeing something very different to everyone else; just compare it to the image of the Deepdale monster produced by Provost Marwick.

The first picture is based on Hutchinson in Tim Dinsdale's *Loch Ness Monster* and is very hard to equate to the image produced by Marwick in the *Orkney Blast*. Here it gets a little weird, if you compare the two sketches to the photo on p.67 you can see a bit of both in the obviously distressed remains; if only it were that simple. There is, of course, a problem with the photograph as there is some doubt as to what the object is, other than the fact that it is a large rotting lump. The photo, found in a German newspaper, masquerading as a picture of the Loch Ness Monster has since been identified as an image of one of the Orkney monsters, but which one it isn't clear, the strong suspicion is that it is actually the Deepdale monster.

For good measure I have also added an image on p.67 of the classic way that a basking shark rots. So what exactly was Hutchinson looking at? It is possible that his take on the Deepdale remains were influenced by his sighting of a sea monster in the Bay of Meil in 1910. However Dinsdale does suggest an alternative reason for the difference, and that is that Hutchinson had actually gone and viewed the wrong remains, turning up at the Hunda site instead. Tempting as this is to believe, Dinsdale does in fairness explain that he was not all that familiar with the location of the Hunda carcass and imagined them as being quite close. While they might not be that far as the hungry seagull flies, Hutchinson would have needed to catch a boat to get to the Hunda corpse.

Are you, like me, wondering how two dirty great big basking sharks came to be washed up on the shore in such a short period of time? Well the answer may very well be entangled with the military defences of Scapa Flow. Dinsdale imagines that one of the pair of creatures swam into a floating contact mine with an all too predictable result considering the vast amount of damage 80 kg of explosives will cause to a ship, let alone a nearby flesh and blood creature. It's an interesting theory, but again without knowing the location of mines it might have been a little hard to be certain of such a fate befalling the two beasts. There is of course a local tradition that suggests that actually the two monsters met their end as a result of a depth charge explosion, either way there was nothing down for these two monsters of the deep in December 1941. While the pair may well have been basking sharks, and in the case of the Hunda corpse a very big basking shark at 30 feet, the reporting of them did lead to the mention of another couple of stories. Sadly only a small sample of these were recorded in a letter written by W. Campbell Brodie to *The Orcadian*. In the letter he states that around half the fishermen operating in Scapa Flow didn't accepted the theory that the remains were both sharks. This in turn led to brief mentions of other sightings. One trawler skipper related a sighting of a long and slender creature with long brownish hair that had risen out of the water and attacked the ship's mizzen mast; the skipper reported that this monster had a long neck and a head like a cow and that the attack happened near to the island of Hoy. Personally I find this tale, like the basking sharks, a little fishy. Maybe a slightly better story is that of David Wylie of Burray who stated that he had seen a creature similar in appearance travelling in the water sound between Burray and South Ronaldsay in 1940/41. If that had been the case there would be little time left for a repeat performance, as in May 1943 the number 4 Churchill barrier sealed this route off forever. So what have we discovered from the mystery carcasses of Orkney? Well for one thing we have learned that rotting lumps are not the easiest of things to identify, also that the decayed remains of a basking shark can be very misleading. We have also learnt that large rotting sea creatures not only attract gulls, but that they really do stink to high heaven.

While it would be tempting to link the Stronsa Monster to the Deepdale and Hunda remains, I think we would be unwise to do so. While it seems likely that the Stronsa Monster was indeed nothing more than a type of shark, the size alone (the same size as the Deepdale and Hunda corpses laid end on end) strongly suggests that something far more interesting than a big basking shark was involved.

Gourock 1942

As we have already seen, during wartime it appears that sea serpents can't help but throw themselves onto the nearest beach. Our next case takes place well away from Orkney on the Firth of Clyde in the town of Gourock, when sometime in the summer of 1942 a very strange beast washed ashore at Cardwell Bay.

With it being summer, and we assume the weather being vaguely warm (it is Scotland after all), it didn't take long for this creature to draw attention to itself; yes like any dead sea creature left in the sun for a few days it wasn't only unfit to go in a sandwich but it was also rather pungent. As a result of this stench a council worker going by the name of Charles Rankin, along with another unnamed council worker headed up along the coast looking for the cause of such an unpleasant aroma. You can only imagine their surprise when they finally reached the cause of the whiff; they weren't faced with a sack of week old herrings but a great big sea serpent carcass. Seeing that this animal was something unusual, Rankin rang the Royal Scottish Museum to see if they were interested in the remains; it appears they were not. Failing with that he considered taking a photo of it but the story goes that as it was wartime, and this stretch of the coast was a restricted area, the Royal Navy refused him permission to get the camera out lest he gave away any secrets that might be in the background. It's not too hard to

Sea Serpent Carcasses—Scotland

believe such restrictions being in place, but it does make the fact that pictures were taken of both the Deepdale and Hunda carcasses all the more remarkable considering that Scapa Flow was home to the British home fleet.

Due to the unhealthy nature of the remains, Rankin was told to dispose of the body. This he did by having it chopped up and buried in the grounds of the municipal incinerator, which in itself doesn't sound all that healthy, and perhaps makes it all the more surprising that today the site of the burial is under the playing surface of the football pitch of St Ninian's Primary School. Fortunately, while the authorities may not have wanted anything to do with the beast, Rankin did make some rather detailed notes on the creature, which give us the following description.

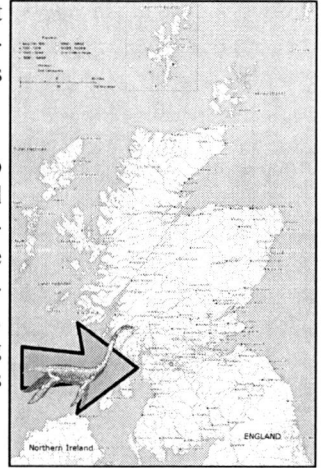

'It was approximately 27-28 feet in length and 5-6 feet in depth at the broadest parts. As it lay on its side, the body appeared to be oval in section but the angle of the flippers in relation to the body suggested that the body section had been round in life. If so, this would reduce

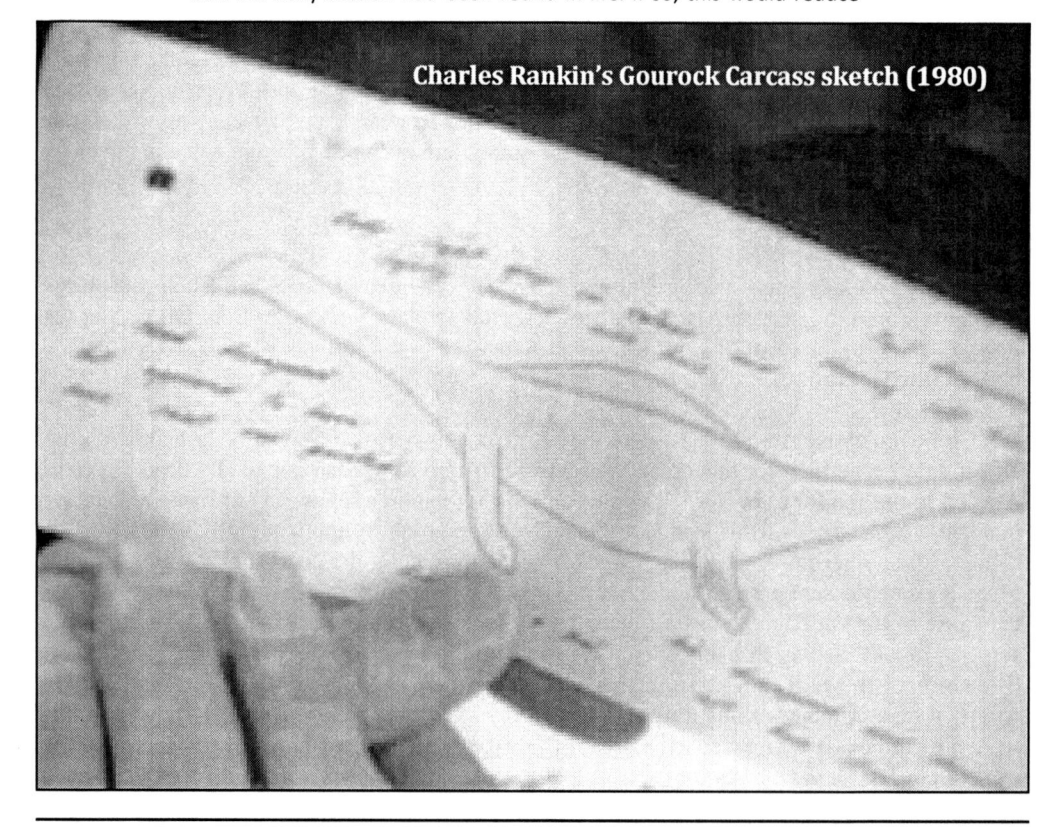

Charles Rankin's Gourock Carcass sketch (1980)

the depth dimension to some extent. With head and neck, the body, and the tail were approximately equal in length, the neck and tail tapering gradually away from the body. There were no fins. The head was comparatively small, of a shape rather like that of a seal, but the snout was much sharper and the top of the head flatter. The jaws came together one over the other and there appeared to be a bump over the eyes—say prominent eyebrows. There were large pointed teeth in each jaw. The eyes were comparatively large, rather like those of a seal but more to the side of the head. The tail was rectangular in shape as it lay and it appeared to have been vertical in life. Showing through the thin skin there were parallel rows of "bones" which had a gristly, flossy, opaque appearance. I had the impression that these "bones" had opened out fan-wise under the thin membrane to form a very effective tail. The tail appeared to be equal in size above and below centre line. At the front of the body there was a pair of "L"-shaped flippers and at the back a similar pair, shorter, but broader. Each terminated in a "bony" structure similar to the tail and no doubt was also capable of being opened out in the same way. The body had over it at fairly close intervals, pointing backwards, hard, bristly "hairs." These were set closer together towards the tail and at the back edge of the flippers. I pulled out one of these bristles from a flipper. It was about 6 inches long and tapered and pointed at each end like a steel knitting needle and rather the thickness of a needle of that size, but slightly more flexible. I kept the bristle in the drawer of my office desk and some time later found that it had dried up in the shape of a coiled spring. The skin of the animal was smooth and when cut was found to be comparatively thin but tough. There appeared to be no bones other than a spinal column. The flesh was uniformly deep pink in colour, was blubbery and difficult to cut or chop. It did not bleed, and it behaved like a thick table jelly under pressure. In what I took to be the stomach of the animal was found a small piece of knitted woollen material as from a cardigan and, stranger still, a small corner of what had been a woven cotton tablecloth—complete with tassels.'

During an interview for the TV show, *Arthur C. Clarke's Mysterious World*, in the episode 'Monsters of the Deep', Charles Rankin goes on to describe the remains as showing no signs of being rotten and it certainly did not resemble the remains of a rotten basking shark. He also gave an indication of the size of the teeth of the beast, describing them as just over an inch in length. It turns out that not all of the beast was buried under the surface of that future football pitch, it seems that Rankin helped himself to one of the 'knitting needle' bristles from the flipper of the creature which he kept in his desk for a number years. It is said that this bristle eventually curled up like a spring, but unfortunately what happened to it after that doesn't seem to have been recorded.

So what creature could the remains have been in life? Rankin provided a sketch of a prehistoric looking creature and appeared to favour a reptile origin for it. The usual basking shark candidate that is often rolled out to explain many a stranding does not fit the reported details. For a start this beast had a mouthful of teeth, something that a basking shark doesn't have. This mouthful of sharp teeth suggests some creature that you wouldn't want to be in the water swimming near when it appears, it was after all far greater in length than the measurement you would expect from the average great white shark. Sadly though, without any pieces to look at, we will never know for sure.

Machrihanish 1944

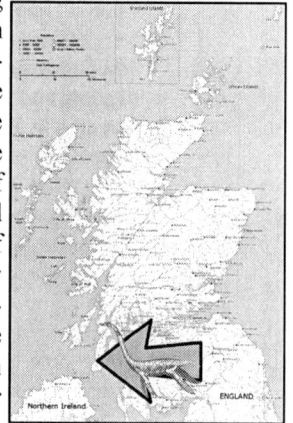

Occurring just a couple of years after the 1942 wave of strandings, the one at Machrihanish is perhaps one of the least known. Once again the sighting took place close to a military base, in this case the RAF base. In recent years RAF Machrihanish has been at the centre of a number of conspiracy theories regarding just what exactly has been going on at the site. Aside from the back engineering of recovered alien spacecraft there has been speculation about the length of the runway. At one time the 10,000 foot long runway was officially the longest in Europe that enabled it to be used by the UK's one time nuclear deterrent, the wonderful Vulcan bomber. However the whispers have it that the long runway was actually for the use of top secret American aircraft that needed somewhere to turn round before heading back home boot full of duty free. The star plane of this mystery group was supposedly the Aurora hypersonic spy plane said to race through the sky at incredible speed before stopping off to refuel at Machrihanish. But why would such a plane need to refuel in Scotland? What possible supply of special aviation fuel could be found at the base? Now I wonder if anyone has ever considered the fact that Scotland does have its own volatile rocket fuel, Buckie; it has certainly been used to fuel an endless string of Neds. Away from the iffy world of conspiracy we get back to Machrihanish, which other than the airport is a small village and a golf course, and back in 1944 I can't imagine it was all that different when a strange creature floated ashore. It has to be said that while reports of the creature spread around the world, these reports were less than forthcoming with specific details, as this piece from a newspaper from Tasmania ably demonstrates.

> *The Mercury* Wednesday 4th October 1944
> Shades Of Loch Ness
> LONDON, Tues. (AAP).
> Washed up from the Atlantic, a strange sea monster more than 20ft. long, described as fur-coated and with enormous eyes and large feet, is attracting large crowds to the Scottish village of Machrhanish, where it lies on the rocks. It resembles neither a whale nor a seal. It is believed to have been killed during naval exercises or action.
> It appears that navel warfare provided the coast of Scotland with a fair number of mystery creatures that needed to be glued back together in able to identify what they might once have been. Shortly after they washed ashore the remains were suspected as being one of a couple of things, either a sea serpent or, rather uniquely, an over-sized polar bear. If it was a polar bear it had done well drifting a long way from its true home in the Arctic north, or of course it could have fallen off a passing ship. But before either theory could make any headway a certain Dr A.C. Stephens from the Natural History Department of the Royal Scottish Museum, after talking to someone on the spot, decided it was nothing more than a basking shark.

Sadly it doesn't appear that anyone has yet managed to track down a photo of this creature, assuming - that is - that a picture was taken of it. Certainly some clarification would be needed to see what the large eyes mentioned in the report actually were, and as for the feet you would be surprised to know just how weird a dead basking shark can look.

Troon 1948

Today, if you mention Troon most folk will only know about its golf course, but back in the 1940s it had its very own sea serpent carcass to look at. From the photos of it, it looked to be on the ripe side; maybe a nose peg would have been needed for a close-up viewing. Much like a good mystery sea monster, the report of this stranding first appeared in 1948 before resurfacing in the mid 1970s, just before a Loch Ness conference, or so an article in the *Glasgow Herald* informed us. The actual article originally appeared in 1948, and was reprinted on 3rd December 1975.

> *Glasgow Herald*
> 'a strange sea monster – it had a head like a horse and looked like a dragon – washed up last night on the beach at Troon, and set the little seaside town agog with excitement.

The 'Dragon of Troon' was torn and gashed, and most of its skin had been washed away – and at some time'

'but no one knew last night what it was – except that it was the weirdest thing they had ever seen. It lay stretched out on the beach between Blackrock and Old Troon Golf Course, and as the rumour spread that 'a monster is here' the town turned out to have a look at it'

'it was about 30ft long, dirty white in colour, but the head was small in relation to the body – which ended with a 10ft long tail.

Its eye sockets dug deep into its forehead. The backbone was complete – and every vertebra was clearly defined.'

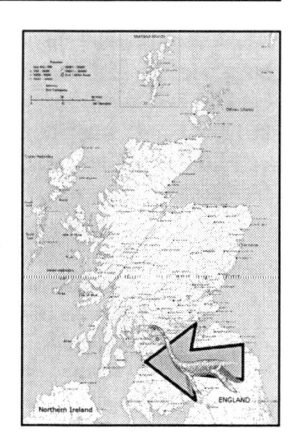

That is basically the article, but there were a few further interesting comments recorded at the time that make interesting reading today.

One witness stated,

A HEAD

TROON'S MONSTER: ANOTHER PHOTOGRAPH.

'it's the nearest thing to the Loch Ness Monster I've ever seen'

It does make you wonder if they really thought that the Loch Ness Monster looked like 30 feet of rotting fish.

Another witness who was interviewed was Mr Ian Fadyen who, the report goes on to tell us, had travelled the seven seas and seen all kinds of sea monsters washed up, but just how many sea monsters that was isn't recorded.

However his view on the Troon Dragon was,

'it's got me beaten. I've just never seen anything like it before.'

From that statement can we really be sure he had actually seen any real monsters whilst travelling the seven seas?

Our last statement came from Sidney Benson from the long lost Calderpark Zoo,

'It might be a bottle-nosed whale.'

Well that was a much better go at identifying the remains, however a basking shark would appear to be a better identification for them.

Irvine 1950s

Just what is it about the Ayrshire coast? It seems that the beaches there are nothing more than magnets for mystery dead critters. We have already had sightings at Prestwick, Troon, and now this one turns up on a beach at Irvine. Sadly the details are rather incomplete but nevertheless they present yet another mystery from the deep.

Due to the nature of the report, at best third hand, it is hard to say how the details may have changed or indeed if the sighting isn't simply a rehashing of one of the other local sightings. However, it's worth having a look at what is known, as the retelling of the details may prompt another witness to appear.

It was some time in the early 1950s that the remains of a creature some 20 foot long washed up on an Ayrshire beach near Irvine. It was described as having a large reptilian head with the rest of the body having a smooth skin. It lacked any signs of having gills, instead it seemed to have an 18 inch 'frill' around its neck. As for its condition, it was said to have shown no sign of decay or damage.

And what happened to this creature? It was said to have been washed out to sea on the next high tide before any experts managed to have a look at it. So this one remains nothing more than a soggy footnote in the sea serpent carcass world.

Girvan 1953

It seems that during the war there was no end of dead critters washing up on the nation's beaches, but once people stopped launching explosives into the water the frequency of strandings being reported once again tailed off.

That was until 1953 when yet another smelly lump drifted ashore on a beach in South Ayrshire.

The remains would come to be known as the Girvan Monster, but the creature actually washed up on the beach in Dipple near the Alginate factory.

For whatever reason the Dipple Monster didn't manage to catch people's imagination, however as the Girvan Monster it appears to have had worldwide appeal as the following two articles from Australian newspapers testify.

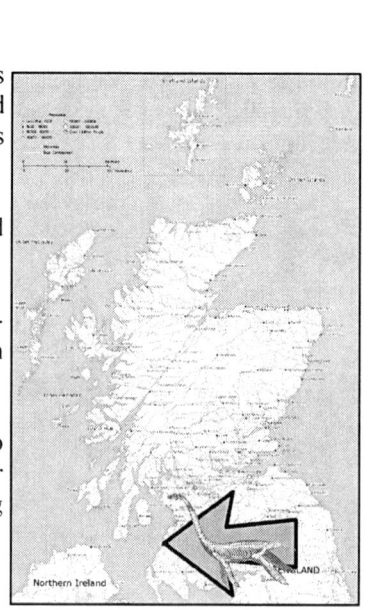

Sea Serpent Carcasses—Scotland

Sunday Herald (Sydney, NSW) Sunday 30th August 1953
New Sea "Monster" Comes To Scotland
From Our Staff Correspondent
LONDON, August 29.-The 5,000 people of the placid fishing village of Girvan, on the coast of Ayrshire, West Scotland, are on the hunt for a prehistoric monster.

They firmly believe they have seen one swimming off shore during the last few days, searching for its lost mate.

The mate, they say, was a grotesque 30ft monster washed up on the beach on August 15. The beast, left by the tide on rocks, had a four-foot "giraffe neck," a camel-shaped head with bone-shielded eyes, four stumpy appendages like legs, and a 12-foot tail.

According to the Provost (Mayor) of Girvan, Mr. D. M. H. Smith, the beast also had "the intestines of a mamal, mon, so it could nae have been ony feesh, ye ken."

BURNT BEAST
All Girvan flocked to gaze at the creature, thrilled by the thought of having acquired a plesiosaurus, a marine reptile active millions of years ago.

The Scottish papers, remembering it was 20 years since the birth of the still elusive Loch Ness monster, made Girvan front page news the

Sea Serpent Carcasses—Scotland

next day, and Fleet Street (London) papers carried stories and pictures.

Experts were hurriedly summoned from Edinburgh, but before they could arrive at Girvan, the people, revolted by the smell, had poured oil on the beast and burnt its whole carcass, except the head.

That had unaccountably vanished almost before their eyes.

From photographs taken before the burning, the thwarted experts decided Girvan's "missing link" was a basking-shark that had died a natural death in the sea, but had been battered by rocks and half eaten by other sea creatures. They returned to Edinburgh disappointed.

But in London, the famous zoologist, Professor Julian Huxley, could not dismiss the affair so lightly. "It was a grave mistake to destroy the creature," he said. "The long neck, tail and legs were most unusual. I would welcome fragments of jaw bones and feet for analysis."

Local people believe they have now solved the mystery of the absent

Girvan stranding, original oil painting by Glen Vaudrey
Oil on stretched canvas measuring 16 x12 inches (40cm x 30cm)

While the Girvan stranding is one of the best described it also resulted in some of the worst photos, so what better way to look anew at this pseudo plesiosaur than with oils.

Sea Serpent Carcasses—Scotland

head - it may be on its way to Professor Huxley under the arm of Tony McTaggart, 60, a wartime liaison officer with General de Gaulle.

HELPED HIMSELF
McTaggart, a tall bearded man in a black Astrakhan hat, is a local recluse, who is reported to have been seen putting the monster's head into a sack on the beach. He is now missing from his cottage.

The Provost, Mr. Smith, also confessed he had helped himself to some of the monster before it was destroyed - a part of the backbone, which he has sent to London for analysis.

This weekend lookouts on the harbour wall and the sloping green hills behind Girvan are maintaining a non-stop vigil for the next appearance of the monster's mate, while the village has become a trippers' paradise.

Thousands are booked to go there by train and road, a local manufacturer is turning out women's scarves with a monster pattern, and outsize potatoes are being sold as "other Girvan monsters."

So-called fragments of the remains will be sold to the highest bidders.

Skipper "Geisha" Sloane, who was the first man to see the strange creature in the Clyde, will run hourly hunts for the monster in his ketch Amethyst, with 12 passengers at 10/ each.

Perhaps we see for the first time a new facet of the aftermath of a stranding, tourism. While you would initially think that 30 foot of rotting basking shark would clear a beach in seconds it appears that in the eyes of some enterprising souls it was a marketing opportunity beyond belief, for while a rotting fish clears a beach, a stranded prehistoric monster fair drags people in. Further confirmation of the marketing potential of a marine corpse is contained in the follow-up article in the *Sunday Herald*.

Sunday Herald (Sydney,NSW) Sunday 13th September 1953
A Village Loved Its Monster
The 5,000 people of Girvan, a placid fishing village on the coast of Ayrshire, Scotland, are mourning the final passing of their monster-and the end to fond hopes of prosperity conjured up by that seven-day wonder.

When the remains of a "fantastic" creature were washed up on Girvan beach last month it was hailed as a survivor from a pre-historic era.

It was "35ft long, covered with horse-like hair, had a 4ft giraffe neck, a camel shaped head, with bone shielded eyes, four stumpy legs, and a 12ft tail."

Rough location of Girvan carcass

"It's a Plesiosaurus," declared the local authorities, meaning the marine reptile that was active millions of years ago.

The Provost (Mayor) of Girvan, Mr. D. H. M. Smith, was quoted as saying: "It has the intestines of a mammal, mon, so it could nae be a feesh, ye ken."

Trippers went to Girvan in their thousands, and villagers began reaping a rich harvest selling souvenirs. Skipper "Geisha" Sloane, a local fisherman, ran hourly hunts for the monster from the local quay in his ketch Amethyst, carrying each time 12 passengers at 10/ each.

Then came experts from Glasgow University, led by young Professor James Parry, Lecturer in Zoology. They took away specimens of the creature.

Townspeople, objecting to the stench on the beach, poured oil on the carcass and set it on fire.

The head had already been purloined by an enterprising Scots hermit who it is reported, has been hawking it around the country. But the experts did not need the head to pronounce judgment. It was nothing more exciting than a basking shark. "We get them in these parts," Professor Parry told "The Sunday Herald." "This one was probably a baby, no more than 15 feet long-certainly not 35ft - that's utterly ridiculous.

"It was almost impossible for any man to identify it because it had been so badly mutilated by rocks and other sea creatures. Local imagination and wishful thinking did the rest."

But Girvan, though in mourning, has a profit to show.

In Sydney last week Mr. G. Whitley, Curator of fishes at the Museum, took a look at a newly arrived photograph of the "prehistoric monster."

Girvan has been identified by some writers as "Elfhame" the traditional kingdom of the fairies and the witches; and there was a supernatural touch about the photograph. After about three seconds Mr. Whitley said: "Without any hesitation, on the evidence of this photograph alone, I can say confidently that these are the remains of a basking shark.

"The object that looks like a skull is the upper part of the cranium. The soft tissue of the long projection you would call the nose has fallen away, and the lower jaw with it.

"The soft tissue of the dorsal fin has disintegrated, leaving fibres which resemble the hair of a horse's mane."

Mr. Whitley said that a similar carcass years ago caused a similar report. He had seen the remains of that carcass, which were preserved in the Museum of the Royal College of Surgeons in London.

The basking shark is a harmless monster, growing up to 40 feet in length, which feeds on the abundant plankton of the Gulf Stream, and is fairly frequently washed up on shores in the north of Scotland.

Meanwhile some suspicion must rest on the inhabitants of Girvan who probably had long envied the tourist popularity of Loch Ness. After the discovery of a "monster" there. Probably a large number of them well know the basking shark, for which an intensive fishery was begun during the war. And the culprits who set the carcass on fire and removed the skull were possibly concerned with something more than the stench.

It certainly looked more like a horrific prehistoric Plesiosaur than any enormous but harmless basking shark. This was the photograph that arrived, in Sydney last week after being used in the west of Scotland to bring thousands of trippers to the village of Girvan.

The traditional reason for the remains being set alight was that the workers in the factory had demanded the remains be destroyed as the smell was very unpleasant; having worked in an office opposite a beach where a large skate washed up I can certainly agree the smell could easily bring tears to your eyes. In what in hindsight looks to be a very optimistic way of removing the remains, oil and petrol were poured on the rotten lump and then it was set alight. Unsurprisingly, considering how much time and energy it takes to cremate anything that size, the blaze failed to destroy the remains. However, if the allegation in the paper is to be believed the fire was not so much an attempt to remove the smell but rather a way of making a basking shark look a little more like a prehistoric sea serpent; as would the removal of the skull, however the idea that someone would think that the skull could be of value is not too hard to believe. But it does make you wonder what they did with it once the monster's 15 minutes of fame were over; I guess it would be too far gone for making soup.

In all it looks like the Girvan Monster was nothing more that the badly treated corpse of a basking shark, so it's perhaps not all that surprising that Girvan never managed to get the tourist trade of monster hunters that Loch Ness continues to attract.

But as luck would have it Girvan does have another claim to questionable fame and that is the Hairy Tree which, according to legend, was planted by the eldest daughter of that most folkloric of cannibals Sawney Bean. It must have been a fast growing tree as when the Bean clan's crimes became known she was hanged from its branches and it was said that occasionally you could hear the sound of a swinging corpse if you stood beneath it. Perhaps the strangest thing about the Hairy Tree is that no one knows where it is today. Now losing a tree is just careless, but saying that, considering there is no proof that any the Bean clan ever existed it may be not that surprising.

Barra 1961

Perhaps it was just the location of the remains on the island of Barra that stopped hordes of tourists heading off to view them. Barra, being an island in the Outer Hebrides, is after all a lot harder to drive to. It's a pity that more folk couldn't manage to arrange a trip out to see it as for once this creature wasn't a rotting basking shark. Saying that, it wasn't a sea serpent either, but perhaps one thing we can say for it was that it smelled a bit.

The initial reports were rather promising, telling of a long-necked carcass, but before the monster hunting bandwagon could start arriving, the case was dealt a fatal blow when Mr Peter Usherwood, a Glasgow University zoologist, flew out to see the remains. It didn't take him

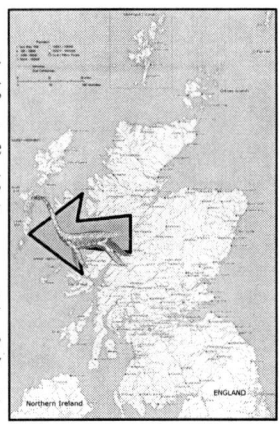

Barra stranding, original oil painting by Glen Vaudrey
Oil on stretched canvas measuring 16 x12 inches (40cm x 30cm)

Capturing the glorious scenery of the charming isle of Barra with the juxtaposed remains of a dead whale.

Sea Serpent Carcasses—Scotland

long to pronounce that the creature was nothing more than what was left of a male beaked whale, the long neck being in fact the backbone of the whale once the flesh had fallen off it and drifted away. Perhaps had the identity of the whale stayed a mystery for a little longer it may well have given some of its basking shark sea serpents a run for their money, but as it was it is little more than a footnote in the world of faux sea serpent carcasses.

North Sea 1963

As we have already seen, sometimes the suspected sea serpent carcasses have a tendency to turn up out at sea, hardly surprising when you consider the creatures' natural habitat. What makes this sighting all the stranger is that the creature spotted was in this instance well away from it natural habitat.

When the Aberdeen trawler *Faithlie* started to pull its net on board the crew could not have imagined what beast was about to break the surface, it was an elephant. Surely that was a headline writer's dream come true, the *Aberdeen Express* came up with 'Aberdeen trawler catches a Jumbo haddock', while across the Atlantic the *Tri-City Herald* was a bit more restrained in its reporting.

> *Tri-City Herald* 31st October 1963
> Scottish Trawler Nets an Elephant

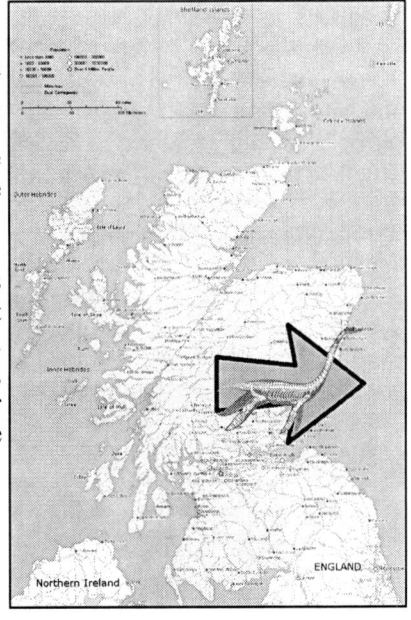

ABERDEEN, Scotland (AP) – This is a fishing tale to top a lot of them - a trawler that caught an elephant.

The crew of the fishing trawler Faithlie returned to port today after a fishing trip in the North Sea and told of hauling up one of their nets with a dead elephant in it.
A spokesman for the trawler's owners said:

'we can only suppose that the elephant either fell from some ship that was taking it to a zoo somewhere, or that, being taken to a zoo, the animal died aboard some ship and was thrown overboard'

The Faithlie dumped the elephant's body back into the sea.

You have to agree that an elephant wouldn't be something you'd be expecting to come across in your nets. But that wasn't the only trawler that caught the soggy pachyderm. It's been said that a trawler from Hull then went and caught it the following week. Now if all that sounds a bit hard to believe then how about this theory for the elephant's origin: the animal is supposed to have been part of a herd on a Russian trawler.

Loch Ness 1972

The Loch Ness Monster has continually managed to capture the imagination of the public for many generations with numerous reported sightings; it has to be said not all these have been as genuine as they first seem. One question that has often been asked is: if there is a monster in the dark waters of the loch why has no one managed to find the remains of any that have died?

We have already seen in 1868 that the odd prankster wasn't beyond throwing a dead beast in the water for the purposes of amusement. What is perhaps more surprising about this next stranding is the length of time between these hoaxes.

When a team of eight English scientists and zoologists arrived up on the shores of Loch Ness in late March 1972, it is hard to say if they honestly believed that they would be making all kinds of news at the start of the next month. If you are in any doubt about that date it would be 1st April 1972, also known as April Fool's Day. Since the early 1930s, the Loch Ness Monster had been drawing scientists and tourists to peer across the dark waters in the hope of solving the mystery, and from then up to the 1970s there had been an endless number of folk who had seen the monster swimming about.

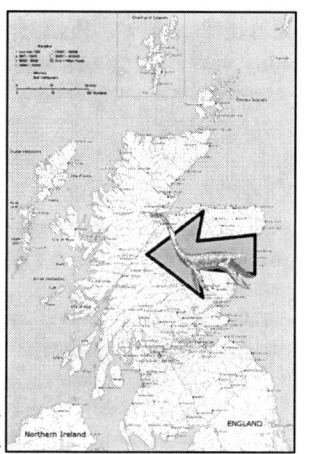

After setting up their base at the *Froyers Hotel* some nine miles from Inverness, the team - headed by Terence O'Brien - had been trying to attract the monster by using a special bait made up of a compound of sex hormones, dried blood, and oatmeal which they dipped in the

Police trying to open the door of a van stopped last night at the Forth Road Bridge.

water. While the team were on the loch side they were approached by police from Inverness who, according to reports, were in the habit of reminding research teams that if they attracted anything with the bait that they should not attempt to damage or remove the creature.

When the subsequent story of the beast's capture made it into the press, it was the *Glasgow Herald* newspaper's comment that best summed up the use of this particular bait.

> *Glasgow Herald* 1st April 1972
> Over the decades our poor Nessie has been hunted by people using everything from mine detectors to submarines and sonic beams. The Yorkshire fellows blithely arrived only last week, equipped with bait of sex hormones, dried blood and oat meal – all of which were sprinkled liberally over the loch. If they have lured a member of the Nessie family to the surface we at least know we have a true and proper monster – one which, is sex-obsessed, blood thirsty and addicted to porridge.

Sea Serpent Carcasses—Scotland

Despite the questionable choice of bait, come 1st April it seemed to have worked. Don Robinson, director of the Flamingo Park Zoo in Scarborough announced that the team had caught the beast, describing it as a green scaly monster weighing in at 1½ ton. The 15 foot long mammal had a scaly body, a massive bear-like head and big protruding teeth. Mr Robinson said that the plan was for the creature to be taken to the zoo where it was planned to put the remains in the deep freezer before rather generously offering the animal to the Natural History Museum.

A local witness, Mr Roderick Mackenzie, a musician (but we can't hold that against him being a credible witness), from Inverness stated that he had seen the monster soon after it landed.

> 'I touched it and put my hand in its mouth'

he added,

> 'it's real, all right.'

He went on to described it as

> 'half bear and half seal, green in colour with a horrific head like a wild bear with flat ears. I was shocked, I had never seen anything like it before.'

While the hotel manageress Mrs Good described the remains as

> 'a dark brown yellowish colour with a big head and teeth and no hair on the face, there were indentations on the back which could be mistaken for humps at between 12 to 16 ft'.

However, on the other side of the loch at the Loch Ness Phenomena Investigation Bureau, a group who had at that time already spent a dozen years trying to identify the monster by various means doubted that the mystery could be solved so quickly. But then it isn't recorded if they had ever tried using bloody porridge as bait. Then again perhaps they just looked at the date on the calendar and knew better.

Perhaps it wasn't surprising that word of this story soon raced around the world, as we can see in this report all the way from the United States where it appeared in the *Pittsburgh Post Gazette*.

> *Pittsburgh Post Gazette* Apr 1st 1972
> Scaly 'thing' found dead in Loch Ness
> INVERNESS, Scotland (AP) – Scottish police intercepted a truck heading for England on this April Fool's eve with a 'green and scaly' creature found dead in Loch Ness, home of the legendry monster.
>
> A team of English zoologists claimed to have found the thing -18 feet long and weighing 1½ tons – off the shore of the Scottish lake yesterday

ALL AN APRIL FOOL'S HOAX

No Monster After All

DUNFERMLINE, Scotland (AP) — A mysterious creature found in Loch Ness, home of Scotland's legendary monster, turned out today to be a frozen bull elephant seal. It was all an April Fool's hoax, police said.

After police stopped English zoologists from sneaking the dead beast out of Scotland, an expert examined the nine-foot, 350-pound creature at Dunfermline police headquarters and found nothing supernatural about it.

Michael Russell, general curator of Edinburgh zoo, said the animal was just a young seal found far from its normal home.

"It is a typical member of its species," Russell said. "It's about three to four years old. . . .

"I have never known them to come near the shore of Great Britain. Their natural habitat is the South Atlantic, Falkland Islands or South Georgia.

"I don't know how long it's been kept in a deep freeze, but this has obviously been done by some human hand."

Police investigating the case, which made front-page headlines in England, speculated that the seal had died aboard a ship bringing a number of this species to a British zoo from the Falkland Islands.

How the seal got from there to murky Loch Ness was another question, however. Possibly, it was dumped overboard in the Atlantic, found by fishermen and turned over to hoaxers, so this theory went.

"It's just an April Fool's Day joke," said Police Superintendent Innis McKay of Inverness. His precinct includes Loch Ness with its never-ending unconfirmed reports of sighted monsters.

Before today's hoax was nailed, a team of English zoologists claimed they fished the creature out of Loch Ness Friday and insisted it was no joke.

---

morning. They bundled it into a small truck and headed for their base at Flamingo Park Zoo in Scarborough, a coastal resort in northeast England.

But the Fifeshire police stopped the truck under a 1933 Act of Parliament prohibiting the removal of 'unidentified creatures' from Loch Ness. The body was taken to nearby Dunfermline for examination.

Some described it as looking like a cross between a seal and a walrus. Don Robinson, director of the Flamingo Park Zoo, said: 'I've always been skeptical about the Loch Ness Monster, but this is definitely a monster, no doubt about that. From the reports I've had no one has ever seen anything like it before….a fishy, scaly body with a massive head and big protruding teeth'

Robinson said members of the team thought the monster has been dead for two or three days. Ever since the early 1930s tales of the Loch Ness Monster have attracted scientists and tourists. Thousands of people over the years have claimed to have seen a scaly creature rising from the depths, flicking a snake like head and rippling a row of humps.

The eight-member zoo team, headed by Terence O'Brien, said a 'large

Sea Serpent Carcasses—Scotland

lump' was seen floating at 9 am yesterday off shore from the Froyers Hotel nine miles from Inverness. The scientists went out in a boat and dragged the creature to shore.

It might be hard to believe but it all resulted in the police chasing down after the poor beast. The investigating team were intercepted by the police near the Forth road bridge. This resulted in four people being held at Dunfermline police station early this morning. A spokesman at the time said:-'No charges have been made so far.' While another spokesmen at police headquarters in Kirkcaldy said: our instructions are not to open the van until the arrival of police from Inverness who are responsible for the inquires' arrangements are being made to identify the creature.

With police intervention it didn't take long for the story to change, as the follow-up in another overseas paper highlights.

The Modesto Bee April 2nd 1972
DUNFERMLINE, Scotland (AP) - The director of an English zoo said a young scientist admitted Saturday that a private joke gloriously misfired and set off a police chase through Scotland in search of the Loch Ness Monster.

Don Robinson, director of the Flamingo Park Zoo in Scarborough, reported that the zoo's education officer , John Shields, had given him a statement saying that he was just trying to hoax a few friends on April Fool's Day – which happens to be Shield's 23rd Birthday.

The joke was to dump a frozen bull elephant seal in Loch Ness for his seven monster hunting colleagues to find, the statement said.

Chased Truck
But it got out of hand on Friday when the team tried to rush their discovery back to the zoo on England's northeast coast. Police chased their truck, stopped it and took the 'monster' to this Fifeshire County town for examination. And there, Saturday morning, two scientists from Edinburgh identified the creature as a big seal brought from the waters of the South Atlantic.

Sheilds, in turn, disclosed that he got the idea for the hoax after hearing about a dead elephant seal brought back recently by an expedition to the Falkland Islands off Argentina, Robinson said. He gained possession of the body and kept it in the deep freeze at another zoo.

Breakfast nearby
On Friday morning the eight man team from the Flamingo Park Zoo was having breakfast at a hotel beside Loch Ness, legendary home of

Sea Serpent Carcasses—Scotland

the monster, about nine miles from Inverness.

The team had been cooperating with the Loch Ness Phenomena Bureau in searching for proof that the monster exists.

At 9 a.m. a passer-by called the team's attention to a body floating about 300 yards offshore. The scientists put out in a boat. They came back dragging with them a creature which was variously described as anything between 12 and 18 feet in length and weighing up to 1½ tons. Some described it has having a bear's head and brown scaly body with claw like fins. Others said it had a green body without scales and was more like a cross between a walrus and a seal.

Exultant Cable
The scientists sent an exultant telegram to their boss, director Robinson, then they loaded the creature, wrapped in blankets, into a truck. After allowing it to be photographed they headed for Scarborough.

The Inverness police, however invoking a 40 year old law prohibiting the removal of 'unidentified creatures' from Loch Ness, asked other police forces to halt the truck.
The Fifeshire, Dunfermline police caught up with the 'monster' and kept it on ice until Michael Rushton, general curator of Edinburgh Zoo, declared it was just a young seal found far from its natural home.

Robinson said he understood that Sheids secretly shipped the seal to Loch Ness and dumped it into the lake in the early hours of Friday, how this was done was unexplained.

So there you go. It was all one big hoax, and as you will soon see it wasn't the last from Loch Ness.

Luce Bay 1981

After the fakery of Loch Ness it's perhaps refreshing to find us once again looking at a real sea serpent carcass, well it's only refreshing if you are standing up wind of it.

The stranding took place at Luce Bay in Wigtownshire in southern Scotland. The bay is some 20 miles wide at its mouth and part of it had been used as a bombing range for the RAF.

It was a hot summer's day in 1981, which almost certainly makes this a smelly one. Stranraer Police received a call from the folk of Sandhead village, which is located at the head of Luce Bay in the Rhins of Galloway.
When Detective Sergeant Brian Park and Detective Constable

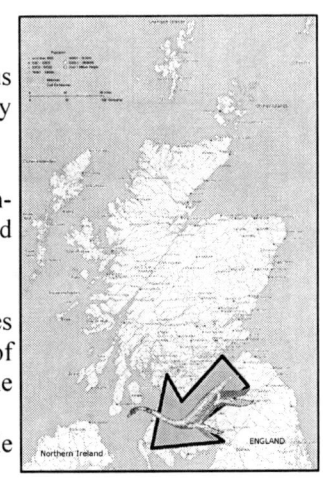

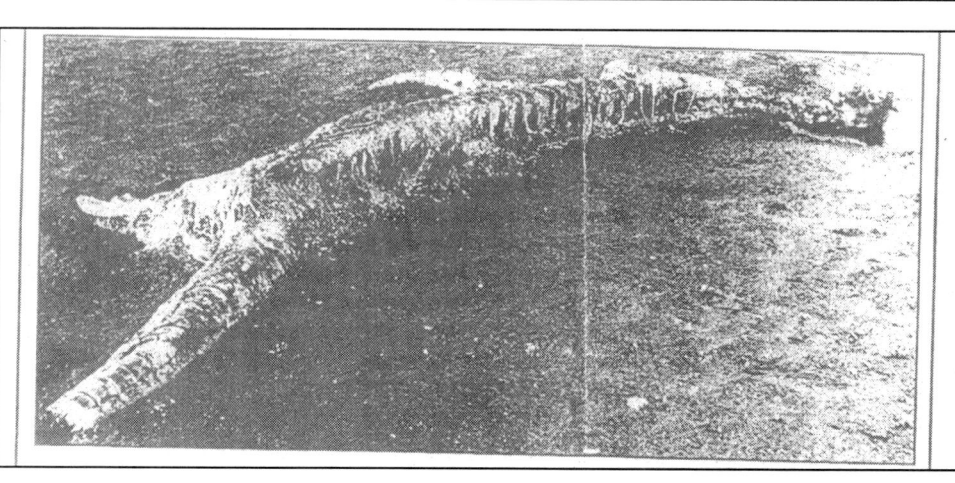

Jim Davitt arrived at the beach they found not only a bad stench but its cause, a large strange lump of decaying animal. What they were looking at were the last remains of a large marine animal that had drifted ashore. Whatever the animal had been in life, in death it could be summed up as 24 feet in length, and it was estimated to have been even longer as the creature appeared to have lost the end of it tail. At the end of a six foot long neck was a head that they described as being shaped like a clenched fist. There was also a powerful pair of flippers at the front, and an exposed ribcage, as well as a set of hind flippers and a tail that was lacking its tip.

The two policemen soon realised that they had something unusual in front of them; they attempted to contact a museum but were told that 'no such creature exists.' And with that ended their interest in the case, which does explain why no one went to look at the remains before they were once again washed out to sea.

Fortunately, before the tide saved the council the job of disposing of the corpse, a couple of photos were taken which, unlike some of the earlier cases looked at, suggest that there was indeed a creature spotted.

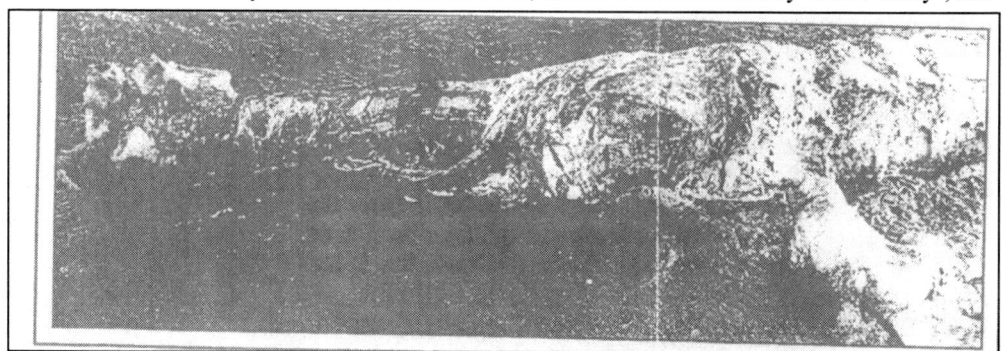

At the time there was speculation that the remains, which were occasionally called 'Lucy', had fallen foul of a practice bombing run and it was as a result of that, that a creature not normally seen so close to the shore found its way into shallow waters and onto the beach. The two police officers were quite certain that the remains were those of just one mystery animal and not a bucket load of old off-cuts dumped at the site to provide a rather smelly hoax.

Benbecula 1990

Over the years, many things have washed up on the beaches of the Western Isles. Within the last four years that has included a couple of sperm whales that have not only become minor attractions but also a source of unbelievable smells, there is after all nothing like forty tons of rotting whale to clear the sinuses, it also attracts a lot of seagulls for some reason. If a washed up carcass fails to float back out to sea during the following couple of days, thus removing its stinking remains to the deep and far away from people, then the mortal remains are put on the back of a low loader and dragged all the way to the council tip just outside Stornoway for burial, and quite possibly a chance to reacquaint itself with some of its new found seagull pals.

Of course, sometimes the things that you find amongst the lines of rotting seaweed don't smell as bad, the empty beer bottles and the pieces of wooden pallet that have escaped from somewhere are not smelly just an eyesore, whereas there are some things that have floated ashore that are a bit more dangerous such as the odd bit of military ordinance that has dropped from a plane or fallen off a ship during one of the many military exercises that the islands get to witness on a regular basis. Every now and then a relic of the First and Second World Wars makes an appearance, such as a great big object washed up looking like a giant spiky ball, so it's best not to kick any of the strange lumps of metal that turn up, and better still don't even think about hitting them with a hammer, you never know what might happen next.

Of course, not all the items that get washed up are so easily identifiable, sometimes a strange lump of rotting creature washes up on the shore and these odd things go by the name of globster. It was Ivan Sanderson, the Scottish cryptozoologist, who in the 1960s gave the name globster to the random lumps of nondescript, putrescent carcasses that wash up on beaches all around the world. So rotten are these lumps of flesh and bone that it is almost impossible to identify of what they are actually the remains. Perhaps they are the last remains of a sea serpent, but then again they could just be the last dregs of a shark that has seen better days, or maybe they are what finally happens to a dead whale that decides that it doesn't want to end up on the council tip. Of course since the widespread use of DNA testing it has become easier to identify what the globster started life as, well that's of course when the result of the testing comes back as something on the database, occasionally it doesn't.

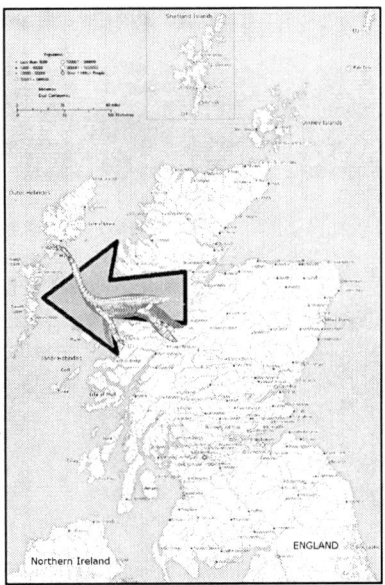

Benbecula globster, original oil painting by Glen Vaudrey
Oil on stretched canvas measuring 16 x12 inches (40cm x 30cm)

These remains had the potential to become one of the greats of cryptozoology, that was until they washed back out to sea and disappeared forever.

Sea Serpent Carcasses—Scotland

With the widespread distribution of globsters it would be unusual for one not to have hit the coastline of the Western Isles and one did just that in 1990. And where did it wash up? Benbecula of course, an island that appears to be a magnet for all things mysterious to wash up on its beaches, from the body of a mermaid to a 27 metre tall silo that no one seems to want to own up to losing. But it was neither a mermaid nor a silo that was the sensation of 1990.

As previously stated not all the things that wash up are easily identified, and this was the case in 1990 when one such mysterious lump washed ashore unobserved onto a beach on the isle of Benbecula. It wasn't long before its remains were discovered by Louise Whitts, a 16-year-old babysitter from Bedlington in Northumbria, while on holiday in the Hebrides. She later described the remains as following.

> 'it had what appeared to be a head at one end, a curved back and seemed to be covered with eaten-away flesh or even furry skin and was about 12 foot long. It smelt absolutely disgusting! But the weird thing was that it had all these shapes like fins along its back. Like a dinosaur or something. We didn't know what it was, although we laughed about it being the Loch Ness Monster.'

An obvious photo opportunity not to be missed, a picture was taken of Louise sitting next to the beast. The picture might have been taken and viewed in 1990, but it would be another six years before the image of Louise and the globster reappeared.

It was only while moving out of her parents' house in 1996 that Louise once again came across the picture at the back of a drawer, and no doubt curious as to what the creature in the picture was, she took the photo along to the Hancock Museum in Newcastle in the hope that it could be identified. If only it were that easy to identify globsters. Alec Coles, curator and keeper of natural sciences at the Hancock Museum, is quoted as saying,

> 'it's not unusual for people to turn up with pictures of mystery animals but almost always we are able to identify them, not in this case though, we haven't seen anything like it before. At first we thought it might be a whale but it doesn't match anything we've got details of although obviously we could be much more certain if we had more pictures or even the body.'

The photos then went on public display that August, but the remains still went unidentified and the carcass would have long since floated back out to sea.

Loch Ness 2001

In recent years, the numbers of suspected sea serpent carcasses washing ashore in Scotland have started to drop off. This isn't so much that creatures were no longer washing ashore but rather the fact that people were starting to recognise them as rotting basking sharks or dead whales. However, in one area of sea serpent remains, sightings have increased, and that is the deliberate hoax. It certainly appears that Loch Ness is not only a location for the budding monster hunter but also for the jolly japes of the hoaxer.

One theory put forward to explain the presence of a lake monster in certain lakes is that the creature is nothing more than an out-sized eel. So imagine the surprise that greeted the discovery of two large eels on the banks of the Loch Ness in 2001. Actually, it might not have been as surprising as it seems, the two dead eels were found on a small beach that was at the foot of some steps heading down from a lay-by on the A82. These two eels were much bigger than the fresh water eels known to be in the loch, but there was a good reason for that and it wasn't that they were a pair of monsters. No, they were a pair of conger eels, which had found their way from coastal waters. A post mortem was carried out on each fish and these revealed that one had been eating mackerel and the other another saltwater fish and, as if further to suggest foul play, both eels had been killed by a puncture to the brain; not natural causes then.

It certainly appears that someone had dropped the eels off by the loch. While there is the chance that they had been thrown overboard from a passing fishing boat it is however more likely that they were dropped off by someone using the lay-by.

Loch Ness 2003

So far, when we have been looking at sea serpent remains they have, for want of a better word, been fresh (that's fresh as in a fly-blown lump of rotting flesh), but this particular set of remains that appeared to have been washed up on the shores of Loch Ness in 2003 had age to them, and we aren't talking months, we are talking years, and not just a decade but millions of years.

One theory about the Loch Ness Monster that seems to be given more coverage than most is that it is a plesiosaur, it also happens to be a completely incorrect one. However, that didn't stop the remains of one of these iconic dinosaurs being discovered on the loch side.

It all appeared to start when Gerald McSorley, a retired scrap dealer from Stirling, tripped and fell into the loch. In the shallow water he found four fossilized vertebrae, well I suppose there are worse things to fall upon. It was actually a Dutch tourist with the group who identified the remains as a fossil.

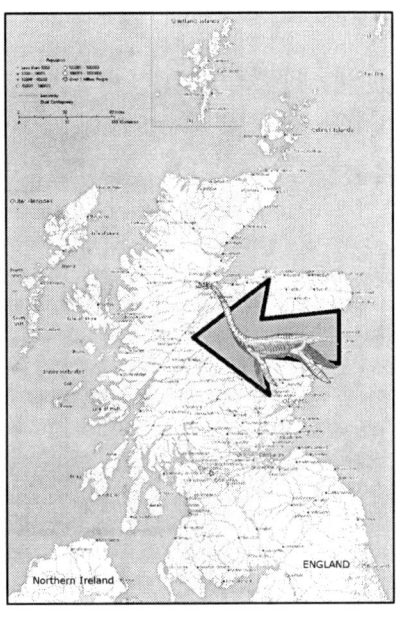

The National Museum of Scotland in Edinburgh confirmed that it was indeed the fossilised vertebrae of a plesiosaur. However, there was strong reason to believe all was not as first seemed. For a start the fossil was embedded in grey Jurassic-aged limestone while the rocks of Loch Ness are much older crystalline, igneous and metamorphic rocks; the nearest place to find a match for this limestone is the Black Isle around 30 miles from Loch Ness. But that wasn't the only problem, the stone on inspection appeared to have been drilled by marine organisms, which strongly suggested that the fossil had been on the seashore until relatively recently.

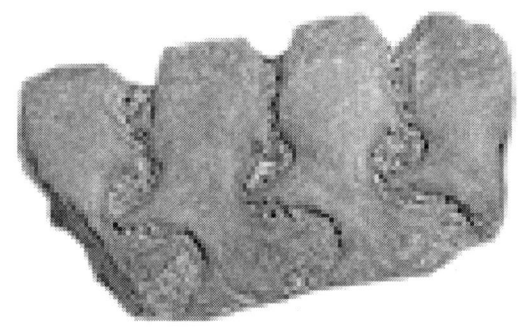

It wasn't looking good for the fossil being that of Nessie, in fact it was looking more likely that the remains had been dumped there by persons unknown.

Whether this was a deliberate hoax or not is still open to debate, certainly the fossil couldn't have found its way there without help, but that doesn't necessarily mean it was left in the water to mislead people. A credible alternative is that apparently at various times tour guides based in the region have used dinosaur bones as props for tour groups and it is possible that this was the case for this particular relic. The fossil could have been used as a demo piece to show what real plesiosaur bones look like and had then been accidently left behind one day.

As for why you would be wasting your time looking for a plesiosaur in Loch Ness, the reason is basically this, they died out millions of years ago while Loch Ness isn't anywhere near as old having been glacially excavated during the last Ice Age around 12,000 years ago.

Loch Ness 2005

Back to Loch Ness again and yet more deliberate fakery. The story goes that two American students were visiting Scotland and, for reasons that will become apparent in a short while, happened to reach the shores of Loch Ness. As luck would have it, they found something while they were there, a big tooth stuck in a deer carcass. Then, as fate would have it, while they were looking at it, a game warden turned up and promptly took it off them, the rotter, but don't worry they managed to get a few photos of it.

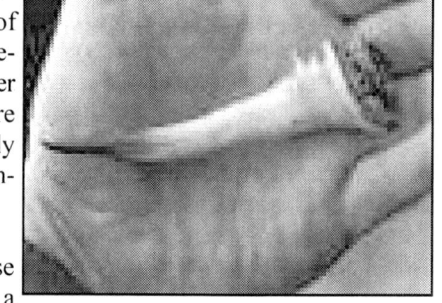

The two students soon set up a website to publicise their find and campaign for its return. However, from a picture of the tooth on the website it didn't take long for folk to identify the tooth as being just the antler of a roe deer. In the end, it all turned out to be a publicity stunt for a horror novel about Loch Ness.

Bridge of Don 2011

You might think that the great days of sea serpent strandings making it into the national press are long gone; amazingly however that's not the case and as recently as 2011 you could find mention of a strange lump of rotting flesh washing up on a beach. The *Daily Mail* reported such an occurrence on 20th July 2011.

> *Daily Mail* 20th July 2011.
> A monster of a find: Couple walking their dogs discover 30ft carcass of sea creature rotting on beach
>
> 'A couple were left shocked when they discovered the rotting body of a sea monster while walking along a beach.
>
> Margaret and Nick Flippence made the incredible find as they exercised their dogs at Bridge of Don, Aberdeen.
>
> Mr Flippence, 59, who lives nearby, said: 'We were stunned. I thought,

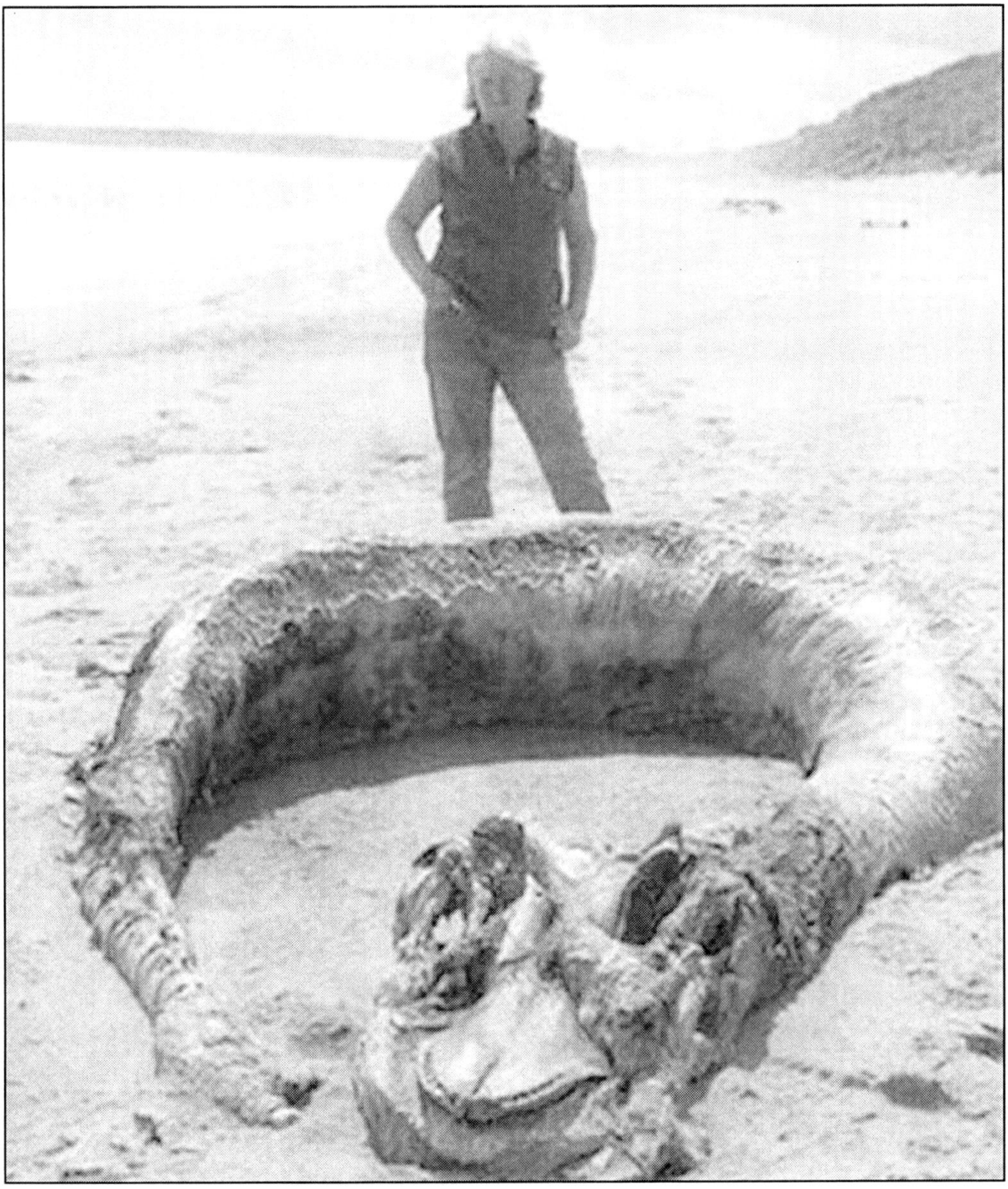

"oh my God what is it?"

'It's like nothing we have ever seen, it almost looks pre-historic,' he told the Sun.

Curled up by the foot of sand dunes was the 30ft-long body of the unidentified animal with head, tail and teeth all discernible.

Experts are now examining the pictures with one suggesting it could be the body of a whale.

A spokesman for the Natural History Museum said: 'We have spoken to one of our mammals curators, and they have confirmed the animal is probably a long-finned pilot whale – *Globicephala melas*.

'Apparently it's not unusual for these to wash up on the shore.'

Rob Deville, a marine life expert at London Zoo, said the body could be that of a killer whale or a smaller pilot whale.

Whale expert Mark Simmonds told the *Sun*: 'it died a long time ago and tides caused the body to wash ashore.'

Unlike many of the past examples, there has been no difficulty in sending images of the remains to various experts for their opinions of what the remains could be. And one thing they all agree on is that it isn't a true mystery sea serpent, but of course that isn't to say that it's not destined to be remembered as one.

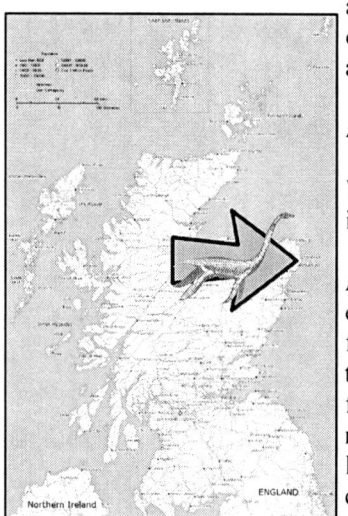

And why is that?

Well, in the age of the internet its picture can be found floating around in all kind of places.

A quick look at some of the comments it has gained in this electronic afterlife turns up a number of interesting identifications, amongst which are dinosaur, frilled shark, prehistoric sea monster, Loch Ness Monster, oarfish, and that old favourite, the plesiosaur. And with that most recent report we reach the end of the list of sightings. Next, we shall have a look at the theories surrounding mystery sea serpent carcasses.

CONCLUSIONS AND RANDOM SEA SERPENT MUSINGS

Over the years, there have been many reports of sea serpent strandings from around the world, which has led to there being enough examples to create a number of categories for the creatures, and these we can use to separate our Scottish sightings into.

Perhaps the most striking of carcasses that wash up are the pseudo plesiosaurs. These remains certainly evoke images of great sea monsters swimming about the oceans. Before we have a look at the examples, it would perhaps be useful to have a look at what we know about the plesiosaur.

In general terms these creatures had a broad body with four flippers and a short tail. Some members of the group possessed a small head on the end of a long neck, and it is that classic plesiosaur shape that a rotting basking shark easily morphs into when it starts to rot.

When you look at a picture of a healthy basking shark it is hard to imagine that when it dies it could take on the shape of the pseudo plesiosaur, but once dead and bits start to fall off there is a noticeable change. As the shark's tissues start to decompose they become soft and as a result parts literally start to drop off. The gills come away and with them the jaw falls off, with the result that nothing remains in front of the front pectoral fin with the exception of the tiny skull and the spinal column still held together by muscle which gives the impression of a long thin neck. While this is happening changes also occur at the other end of the body, as parts decompose and float off it starts to take on the shape of a long thin tail. If changing body shape wasn't enough, there is also a change to the surface texture of the deceased shark. The fibres of the surface muscles start to break up and take on the appearance of whiskers when the skin rots or is eaten away, and this then starts to give the fish a coat of what looks for all the world like stiff fur; its colour can vary from a dirty white to a reddish shade as the body decomposes further.

Of course just because people keep finding pseudo plesiosaurs doesn't mean that there aren't still people claiming that the real live plesiosaurs are waiting to be found swimming in the world's oceans.

Sea Serpent Carcasses—Scotland

Scotland's shores have certainly had their fair share of pseudo plesiosaurs with the beaches of Orkney playing host to some of the finest.

But not everything that floats ashore is a pseudo plesiosaur, there is another worldwide mystery lump and that is the globster. This rather lovely descriptive term describes an unidentified organic mass that has no visible eye, no defined head and no apparent bone structure. There are a couple of things that you can say about globsters, one is that they are not that easy for the average observer to identify, and secondly they do smell.

It appears that the globster is nothing more than a great lump of whale blubber that after death has come adrift from the bones of a dead whale. This buoyant mass then floats off and occasionally washes up on some beach. Easily the best example of a globster to come ashore floated onto a beach in Benbecula in 1990.

Occasionally something washes ashore that is neither a globster nor pseudo plesiosaur; I shall refer to these critters as faux sea serpents. These animals or inert objects appear that unusual to the observer that they appear to be the very remains of a sea serpent; unlike a deliberate hoax they are not a result of mischief but rather misidentification. The reasons for this can be many; in this book two very different examples are covered. In 1872 we have a log that did a very good impression of being a sea serpent when at sea, but turned out to be a very wooden actor when on land. While in the Firth of Forth we have the crew of the *Sovereign* hauling aboard an oarfish, a fish they claim was too foul to feed to Scandinavian dogs, but a fish that today is still mistaken for a sea serpent on the odd occasion it washes ashore.

Of course the most troublesome sea serpent carcass is the hoax. As we have seen in the reports so far, there are a good number of hoaxes to have been perpetrated on an unsuspecting public. It does appear that Loch Ness has had more than its fair share with a track record of all kinds of dead critters being rudely dumped in the area with nothing more than the aim of misleading the unwary.

So with all those possible reasons for the carcass being anything but a sea serpent has there been any true sea serpent remains washed up on the shores of Scotland? Well, the answer has to be that so far it doesn't appear that any of the cases we've looked at could firmly be classed as a true sea serpent. This isn't to say that there are not some cases that seem to pose more questions than they answer. We shall have a look at some of the intriguing cases.

The giant woman of Alba
Sadly the passage of time means that we will never know what the reportedly giant woman who washed ashore in 906 actually was.

Stronsa Monster
While we have plenty of information on the Stronsa Monster, and we even have parts of it in jars, there are plenty of mysteries still to be answered. For instance, what happened to the missing parts, one of which is described as the brain, while another missing piece was described at

the time as a paw? It wasn't until the 1940s that these two pieces disappeared in the Blitz as the story goes, but does that mean that they were destroyed in some fiery conflagration or simply mislabelled and are, even as I write this, collecting dust in the back room of a museum? Prior to their disappearance they were said to be in the collection of Sir Everard Holme (or Holm) and the Royal Society.

Another missing piece didn't so much disappear in a bang but seems to have simply vanished into thin air. At one time it was known that an item described as a framed piece of skin which, according to its last known description, was grey in colour, was to be found hanging on the wall of Tankerness Hall in Orkney before finding its way to Stromness Museum. Sadly contact with that museum has failed to bring it to light.

At a reported 55 feet in length, the Stronsa Monster would easily be the biggest basking shark known. Could it be an example of a large mystery fish occasionally sighted in the waters off Shetland and referred to locally as a 'brigdie' which it appears was a type of large basking shark?

Hunda and Machrihanish

The quest continues to find photographs of these two creatures that washed up during the Second World War. These would obviously make the work of identifying the creature all the easier. Within recent years a number of pictures of sea serpent carcasses have been found that have helped to clear up several ongoing cases. The German cryptozoologist Markus Hemmler has been doing sterling work uncovering the photo of the Deepdale carcass that features in this book and in 2010 he found the photos that answered one of cryptozoology's greatest mysteries: the case of Trunko the globster. Trunko is the truly shockingly bad name for a globster that had washed up on a South African beach in 1924. At the time of its stranding Trunko was described as being a fish-like polar bear. It may be of interest to compare that description to that of the Machrihanish report of a polar bear.

While it is not known if there is a picture of the Machrihanish beast out there, there are rumours of a photo of the Hunda beast, a description of which the picture in this book has been based on. Who knows, it may not be too long before Markus Hemmler tracks them down.

Gourock and Prestwick

It is perhaps the creatures reportedly found at Gourock and Prestwick in the 1940s that raise the most unanswered questions.

Both of these creatures differ greatly from the other sightings, for a start neither creature can easily be solved with the well-used rotting basking shark carcass answer. These two animals are closer to what at the time was understood to be a plesiosaur. Another thing that appears to be the case with these two critters is that before the 1970s there doesn't seem to be any sign of them recorded anywhere.

It was on 25th April 1977 that the Japanese fishing boat the *Zuiyo Maru* hauled something unpleasant out of the water some 30 miles off the New Zealand coast. It was roughly 33 feet long and had certainly seen better days, in fact it smelt so bad the crew soon threw it back overboard, but not before they took some photos of it. For all the world it looked like a dead plesiosaur had been caught in the nets. It was some time before remains of it were tested and it turned out to be nothing more than a badly decomposed shark, so it turns out it was just a pseudo plesiosaur. However that doesn't mean it didn't launch a wave of speculation of surviving plesiosaurs in the world's oceans, and it is in the light of this event that our two Scottish creatures come out of the shadows.

The Prestwick carcass first appears in a letter dated 6th June 1977, while the Gourock creature makes its first appearance as part of *Arthur C Clarke's Mysterious World* television series. Research for this show was started in 1978, and sadly no one can recall where they first came across the witness Charles Rankin.

So far I have been unable to find any mention in the newspapers of the time for either sighting, that is not to say that somewhere out there in some local paper archive the proof exists. However I do have to raise some questions that should be asked.

In the Gourock case the witness Charles Rankin is recorded as having said that he was refused permission to take any photos of it. If we compare this with Scotland's other wartime sea serpent carcasses we find a vastly different story. Machrihanish was a military site and while there is no known photograph it didn't stop reports of it making the news nationwide.

While in Orkney, by the time of the Deepdale and Hunda strandings Scapa Flow was one of the most protected areas in the country being the base of the home fleet, while the Orkney islands themselves played host to a garrison of 10,000 troops who manned the various defensives. Despite all the security restrictions it didn't stop any of the reports from making the national news, nor did it stop the taking of pictures of the creatures even though it appears that the location where the Hunda beast washed up was in the vicinity of the anchor point for the anti-torpedo net. Yet Gourock, for reasons unknown, appeared to be far more sensitive to prying eyes.

There is also the mystery of Charles Rankin's sketch of the Gourock remains. Certainly they could be taken to be those of a pseudo plesiosaur, that is the dregs of a rotting basking shark. Yet he maintained that the remains were, for want of a better word, fresh. The sketch also resembles a drawing of the reported lake Khaiyr monster from Russia, a creature first reported in 1964. It is possible therefore that by the time that the witness was being interviewed for the *Mysterious World* programme, his memory of the events nearly 40 years previous could have been influenced by this picture.

With the Prestwick creature we are faced with the similar problems of a lack of corroborating evidence from the time. As no mention of trying to take a photograph of the creature is recorded we are again left with just a description that could well have been influenced during the 30 years between the sighting and the letter recording the experience.

But before we totally dismiss these two sighting we should also consider that back in wartime, unlike today, the availability of photographic equipment was not readily to hand, whereas today nearly every hand held electronic device can take a picture.

There is also one other intriguing possibility, that is that the creatures mentioned at both Gourock and Prestwick could have influenced the memories of each sighting over time. Certainly the two locations are only separated by around 34 miles so close enough for word to get around. While the measurements given for the two remains differ considerably, other details do suggest a possible link, as for the descriptions both animals are described as having a mouthful of teeth, a long neck, long tail and four flippers.

But unless new witnesses, a photograph, or more importantly the last mortal remains of these creatures come to light sometime in the future then we are unlikely ever to know what exactly did wash up in those wartime years.

One last thing

Finally, it is worth remembering that the days of sea serpent carcasses washing ashore are not over, and who knows what might be coming ashore after the next storm. Just remember one important point, it is always better to stand upwind of any carcass on the beach.

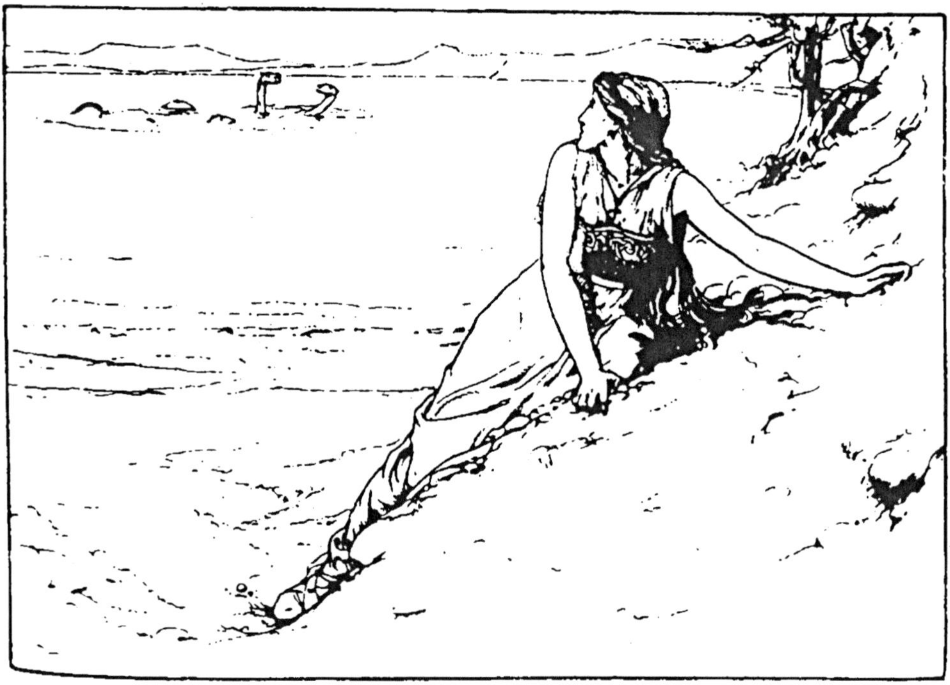

CHRONOLOGY OF SIGHTINGS AND RELATED EVENTS

906 Alba
Something very large and mysterious washed up that years later Irish monks would record as being a large woman.

1808 Hebrides
The Reverend Donald MacLean observes a sea serpent of the isle of Coll in the Hebrides, which he would later state to be the Stronsa Monster. As that creature would turn out to be a basking shark it leaves the question open, what had the good Rev. actually seen?

1808 Stronsa
In the aftermath of a violent storm the remains of a 55 foot long animal are washed ashore. At first it was thought to be the great sea serpent, these days it is now recognised as being a large basking shark.

1810 Scalloway
A mystery creature was described as resembling some vessel turned upside down floated in the bay for two weeks before disappearing. No one was brave enough to go out and get a closer look.

1821 Stornoway
A very large eel was pulled from the waters of a Hebridean loch.

1822 Caledonian Canal opens connecting Loch Ness to the sea.

1830 Benbecula
The body of a mermaid washes up after supposedly being hit on the head with a stone thrown by a Ned.

1848 The crew of *HMS Daedalus* spot a sea serpent causing renewed interest in sea serpents.

Sea Serpent Carcasses—Scotland

1848 Firth of Forth
A fishing boat lands an oarfish amongst its catch.

1849 Usan
The crew of a fishing boat land a marine worm, which they think is a juvenile sea serpent.

1851 Griais
A sea serpent having an itch on the rocky shore leaves a few scales behind.

1868 Loch Ness
Some cheeky scamster dumps a dead northern bottlenose whale in the Loch.

1872 Isle of Man
People pay good money to view a log that looks like a sea serpent. As they say, there's one born every minute.

1877 Oban
A shocking fake tale of an epic struggle to land a sea serpent.

1894 Orkney
The rotting remains of something strange come ashore - it had 'terrible teeth'.

1899 Caledonian Canal
While the locks were being drained a creature looking a little like an eel with a long mane was spotted.

1908 North Atlantic Ocean
The rotting remains of a whale end up in a fishing net, sparking reports of a sea serpent.

1933 Loch Ness
The Loch Ness Monster makes its first modern appearance, causing a rise in reported sightings of stranded sea serpents that might just be the Monster.

1934 Dunnet Sands
The remains of a 29 foot long animal wash ashore. It doesn't take long for those remains to be connected to sightings of both the Loch Ness Monster and a sea serpent off Cape Wrath.

1934 Moray Firth
Yet another oarfish landed.

1939
Start of the Second World War, which among other things leads to a rise in sea serpent strandings.

Sea Serpent Carcasses—Scotland

1939-45 Prestwick
During the war years something similar to a plesiosaur drifts ashore.

1941 Deepdale
The rotting remains of a basking shark wash ashore giving the world the *Scapasaurus*.

1942 Hunda
The second *Scapasaurus* washes ashore. This also is the rotting remains of a basking shark.

1942 Gourock
Something very unusual washes up; a witness to the event gives a description of a creature very similar to that of a plesiosaur.

1944 Machrihanish
According to some, a giant polar bear washes up, to others it's a basking shark that is past its sell by date.

1948 Troon
The Dragon of Troon washes ashore, once again it looks to be the remains of yet another basking shark.

1950s Irvine
A strange eel-like creature washes ashore.

1953 Girvan
Despite the number of day-trippers who went to look at this rotting lump it was once again a basking shark.

1961 Barra
There is little excitement when a sea serpent washes ashore, and even less when it turns out to be a male beaked whale.

1963 North Sea
In a rather surreal moment an elephant ends up in a fishing net.

1972 Loch Ness
After a break of many years, the first in a new wave of Loch Ness hoax monster corpses; this time it's a dead elephant seal.

1981 Luce Bay
Two policemen find what appears to be the last remains of yet another basking shark.

1990 Benbecula
A rather splendid looking globster washes ashore on a lonely beach.

Sea Serpent Carcasses—Scotland

2001 Loch Ness
Yet another hoax when a couple of conger eels are dumped on the foreshore of the loch.

2003 Loch Ness
At long last we have a confirmed trace of a plesiosaur, sadly it's another hoax, this time using some fossils.

2005 Loch Ness
The third Loch Ness hoax of the new century, this time a deer antler is held up to be a tooth of the monster.

2011 Bridge of Don
A dead whale gets a few headlines as it becomes the latest sea serpent carcass to make it into the printed press.

BIBLIOGRAPHY

Ancient Monuments of Orkney (Historic Scotland, 1989)
Arnold, Neil *Monster The A-Z of Zooform Phenomena* (CFZ Press, 2007)
Bright, Michael *There are giants in the sea* (Robson Books, 1989)
Coleman, Loren & Huyghe, Patrick *The field guide to Bigfoot & other mystery primates* (Anomalist Books, San Antonio, 2006)
Costello, Peter *In Search of Lake Monsters* (Garnstone Press, 1974)
Dennison, Walter Traill *Orkney Folklore and Sea Legends* (The Orkney Press, 1995)
Dinsdale, Tim *Loch Ness Monster* (Chilton Company book division, Philadelphia, 1962)
Dinsdale, Tim *The Leviathans* (Routledge & Kegan Paul, London, 1966)
Eberhart, George M. *Mysterious Creatures* (ABC-CLIO, Inc., 2002)
Freeman, Richard *Dragons more than a myth* (CFZ Press, 2005)
Fuller, Errol *The Great Auk* (Errol Fuller, 1999)
Harrison, Paul *Sea Serpents and Lake Monsters of the British Isles* (Robert Hale, 2002)
Heuvelmans, Bernard *In the wake of the sea serpents* (Hill & Wang, 1968)
Konstam, Angus *Scapa Flow The defences of Britain's great fleet anchorage* (Osprey Publishing, 2009)
Marwick, Ernest *The folklore of Orkney and Shetland* (Birlinn, Edinburgh, 2000)
Meurger, Michel *Lake monster traditions a cross-cultural analysis* (Fortean Tomes, London, 1988)
Muir, Tom *The Mermaid Bride and other folk tales* (Kirkwall Press, 1998)
Newton, Michael *Encyclopedia of Cryptozoology a global guide* (Macfarland & Company inc, Jefferson, North Carolina and London, 2005)
Oudemans, A.C. *The great sea serpent* (Cosimo Classics, New York, 2007)
Pálsson, Herman & Edwards, Paul *Orkneyinga Saga translated* (Penguin, London, 1981)
Reader's Digest *Folklore, Myths and Legends* (Reader's Digest, 1973)
Reeves, Randall R., Stewart, Brent S., Clapham, Phillip J., Powell, James A. *Sea Mammals of the world* (A & C Black, London, 2002)
Shuker, Dr Karl P.N. *Dr Shuker's Casebook in pursuit of marvels and mysteries* (CFZ Press, 2008)
Shuker, Dr Karl P.N. *In Search of Prehistoric Survivors* (Blandford, London, 1995)
Shuker, Dr Karl P.N. *The beasts that hide from man, seeking the world's last undiscovered*

animals (Paraview Press, New York, 2003)
Vaudrey, Glen *Mystery Animals of the British Isles: The Northern Isles* (CFZ Press, 2011)
Vaudrey, Glen *Mystery Animals of the British Isles: The Western Isles* (CFZ Press, 2009)
Watson, Ronald *The Water Horse of Loch Ness* (2011)
Welfare, Simon & Fairley, John *Arthur C. Clarke's Mysterious World* (Fontana, 1980)
Wickham Jones, Caroline *Orkney A historical Guide* (Birlinn, Edinburgh, 1998)
Woodley, Michael A *In the Wake of Bernard Heuvelmans* (CFZ Press, 2008)
Zell-Ravenheart, Oberon & DeKirk, Ash *'LeopardDancer' A wizards bestiary* (The Career Press inc, 2007)

A

Alba 11, 108, 113
Angus 44
antennae 70, 72, 74
Architeuthis monachus 37
Arthur C. Clarke's Mysterious World (TV show) 78, 110
Ayrshire
  Girvan 82–89, 115
  Irvine 82, 115
  Prestwick 61–63, 109, 110–111, 114
  Troon 79–81, 115

B

Barra 89–91, 115
basking sharks 78, 79, 107, 109, 113
  Deepdale 66, 69, 115
  Girvan 84, 88, 89
  Hunda 72, 74, 75, 115
  Orkney 54
  Stronsa 24–29, 30, 33, 34–35, 115
  Troon 81
Bay of Fundy 35
Bay of Meil 66, 67
beaked whale 91, 115
Beale, Dr Yvonne 34
Bean, Sawney 89
Beast of Hunda *see* Hunda
Benbecula 39–43, 99–102, 113, 115
Bogha mem Crann 42
bottlenose whales 48, 81, 114
Bridge of Don 104–106, 116
brigdie 109
bullhead 58, 59
Burray 71, 75

C

Caledonian Canal 54–55, 113, 114
Canada 35
Cape Wrath 57–59
Carcharodon megalodon 35
Cardwell Bay 75–78
Chimaera 31
Chimaera monstrosa 31
Coll 33, 113
conger eels 39, 102–103, 115
Corpach 54–55
Cuile 41, 42

D

Daedalus, HMS 113
Deepdale 63–71, 73, 74, 110, 115
deer antler 104
Dinsdale, Tim 69, 74
DNA testing 34, 99

Dragon of Troon 79–81, 115
Drinnon, Dale 61
Dunnet Sands 57–59, 114

E

eels 39, 54–55, 102–103, 113, 114, 115
elephant seal 96, 115
elephants 91–92, 115

F

Faithlie (boat) 91–92
fakes *see* hoaxes
faux sea serpents 91, 108
Findhorn 60
Firth of Clyde 75
Firth of Forth 43, 113–114
Florida 12
Forth River 43
fossils 103–104
frilled shark 106

G

giant octopus 12
giant squid 11, 36, 37
giant woman 11, 108
Girvan 82–89, 115
Globicephala melas 106
globsters 99–102, 108, 109, 115
Gordius marinus montagu 44
Gourock 75–78, 109, 110–111, 115
Greenland 16
Griais 46–47, 114

H

Hairy Tree 89
Halsydrus pontoppidani 18
Hebrides 113
    Barra 89–91, 115
    Benbecula 39–43, 99–102, 113, 115
    Lewis 38–39, 46–47
    Stornoway 38–39, 99, 113
Hemmler, Markus 109
hoaxes 48, 53, 92–97, 102–103, 104, 108, 115–116
Hoy 75
Hunda 71–75, 109, 110, 115

H

Ichthyopterygia 70
Irvine 82, 115
Isle of Man 48–49, 114

Khaiyr, lake 110
killer whale 106
Kirkwall 54
kraken 35–37

L

lake monsters 38–39, 102
    see also Loch Ness Monster
Lamna cornubica 33
Lewis 38–39, 46–47
Lineus longissimus 44
Loch Ness 55, 57–58, 103, 113
Loch Ness Monster 47–48, 59–60, 74, 81, 106, 114
    hoaxes 48, 92–97, 102–103, 104, 108, 115–116
    surgeon's photograph 57, 60

log of wood 10, 49, 108
Luce Bay 97–99, 115

M

Machrihanish 78–79, 109, 110, 115
mammoth teeth 57
marine reptiles 70, 87
marine saurian 70
marine worms 44, 114
McCaig's Tower 49–50
mermaids 39–43, 113
miller's thumb 58, 59
Moray Firth 59–61, 114

N

Nessie *see* Loch Ness Monster
New Zealand 110
Newfoundland 37
North Atlantic Ocean 55–56, 114
North Sea 91–92, 115
Nunton 41, 42

O

oarfish 43, 60–61, 106, 108, 114
Oban 49–52, 114
octopus 12
Orkney 12, 53–54, 110, 114
   Deepdale 63–71, 73, 74, 110, 115
   Hunda 71–75, 109, 110, 115
   Stronsa 12–35, 75, 108–109, 113

P

pilot whales 106
pinniped 54

plesiosaurs 71, 106, 107, 109–110, 116
   Girvan 83, 87, 89
   Gourock 115
   Loch Ness 103–104
   Prestwick 114
   pseudo 107–108, 110
   Stronsa 34
polar bears 79, 109, 115
porbeagle shark 33
Prestwick 61–63, 109, 110–111, 114
pseudo plesiosaurs 107–108, 110

R

RAF Machrihanish 78
Rankin, Charles 75–78, 110
reptiles 63, 70, 78, 82, 87
Russia 110

S

Sanderson, Ivan 99
Sandhead 97
Scalloway 35–37, 113
Scapa Flow 64–65, 71, 74–75, 110
Scapasaurus 70, 71, 72, 73, 115
sea reptile 72
   see also reptiles
sea snake 15, 32
seals 96, 97, 115
Selache maxima 33
sharks 32–33, 34–35, 70, 73, 78, 106, 110
   see also basking sharks
Shetland 35–37, 109
South Africa 109
South Ronaldsay 75
Sovereign (boat) 43
sperm whales 99

Squalus maximus 24–29, 30
  see also basking sharks
squid 11, 36, 37
St Augustine Monster 12
St Columba 47
St Margaret's Hope 72, 73
St Ninian's Primary School 76
Stornoway 38–39, 99, 113
Stronsa Monster 12–35, 75, 108–109, 113
Sucik, Nick 42

T

Thimble Tickle Bay 37
tourism 59, 86, 89, 104
Traittland Ronsay 53, 54
Troon 79–81, 115
Trunko 109

U

Usan 44, 114

W

walrus 53, 97
wartime 99, 110, 111, 114–115
  Gourock 75–78, 109, 110–111, 115
  Hunda 71–75
  Machrihanish 78–79, 110
  Orkney 63–71
  Prestwick 61–63
Welfare, Adam 42
Western Isles
  Barra 89–91, 115
  Benbecula 39–43, 99–102, 113, 115
  Lewis 38–39, 46–47
  Stornoway 38–39, 99, 113

Whale Firth 37
whale sharks 35
whales 11, 12, 16, 55, 102, 108, 114
  beaked 91
  bottlenose 48, 81, 114
  killer 106
  pilot 106
  sperm 99
Whammond, William 61–63
Wigtownshire 97
wood, piece of 10, 49, 108

Y

Yell 37

Z

Zuiyo Maru (boat) 109–110

STILL ON THE TRACK OF UNKNOWN ANIMALS

The Centre for Fortean Zoology, or CFZ, is a non profit-making organisation founded in 1992 with the aim of being a clearing house for information, and coordinating research into mystery animals around the world.

We also study out of place animals, rare and aberrant animal behaviour, and Zooform Phenomena; little-understood "things" that appear to be animals, but which are in fact nothing of the sort, and not even alive (at least in the way we understand the term).

Not only are we the biggest organisation of our type in the world, but - or so we like to think - we are the best. We are certainly the only truly global cryptozoological research organisation, and we carry out our investigations using a strictly scientific set of guidelines. We are expanding all the time and looking to recruit new members to help us in our research into mysterious animals and strange creatures across the globe.

Why should you join us? Because, if you are genuinely interested in trying to solve the last great mysteries of Mother Nature, there is nobody better than us with whom to do it.

Members get a four-issue subscription to our journal *Animals & Men*. Each issue contains nearly 100 pages packed with news, articles, letters, research papers, field reports, and even a gossip column! The magazine is Royal Octavo in format with a full colour cover. You also have access to one of the world's largest collections of resource material dealing with cryptozoology and allied disciplines, and people from the CFZ membership regularly take part in fieldwork and expeditions around the world.

The CFZ is managed by a three-man board of trustees, with a non-profit making trust registered with HM Government Stamp Office. The board of trustees is supported by a Permanent Directorate of full and part-time staff, and advised by a Consultancy Board of specialists - many of whom are world-renowned experts in their particular field. We have regional representatives across the UK, the USA, and many other parts of the world, and are affiliated with other organisations whose aims and protocols mirror our own.

You'll find that the people at the CFZ are friendly and approachable. We have a thriving forum on the website which is the hub of an ever-growing electronic community. You will soon find your feet. Many members of the CFZ Permanent Directorate started off as ordinary members, and now work full-time chasing monsters around the world.

Write to us, e-mail us, or telephone us. The list of future projects on the website is not exhaustive. If you have a good idea for an investigation, please tell us. We may well be able to help.

We are always looking for volunteers to join us. If you see a project that interests you, do not hesitate to get in touch with us. Under certain circumstances we can help provide funding for your trip. If you look on the future projects section of the website, you can see some of the projects that we have pencilled in for the next few years.

In 2003 and 2004 we sent three-man expeditions to Sumatra looking for Orang-Pendek - a semi-legendary bipedal ape. The same three went to Mongolia in 2005. All three members started off merely subscribers to the CFZ magazine. Next time it could be you!

We have no magic sources of income. All our funds come from donations, membership fees, and sales of our publications and merchandise. We are always looking for corporate sponsorship, and other sources of revenue. If you have any ideas for fund-raising please let us know. However, unlike other cryptozoological organisations in the past, we do not live in an intellectual ivory tower. We are not afraid to get our hands dirty, and furthermore we are not one of those organisations where the membership have to raise money so that a privileged few can go on expensive foreign trips. Our research teams, both in the UK and abroad, consist of a mixture of experienced and inexperienced personnel. We are truly a community, and work on the premise that the benefits of CFZ membership are open to all.

Reports of our investigations are published on our website as soon as they are available. Preliminary reports are posted within days of the project finishing.

Each year we publish a 200 page yearbook containing research papers and expedition reports too long to be printed in the journal. We freely circulate our information to anybody who asks for it.

We have a thriving YouTube channel, CFZtv, which has well over two hundred self-made documentaries, lecture appearances, and episodes of our monthly webTV show. We have a daily online magazine, which has over a million hits each year.

Each year since 2000 we have held our annual convention - the Weird Weekend. It is three days of lectures, workshops, and excursions. But most importantly it is a chance for members of the CFZ to meet each other, and to talk with the members of the permanent directorate in a relaxed and informal setting and preferably with a pint of beer in one hand. Since 2006 - the Weird Weekend has been bigger and better and held on the third weekend in August in the idyllic rural location of Woolsery in North Devon.

Since relocating to North Devon in 2005 we have become ever more closely involved with other community organisations, and we hope that this trend will continue. We have also worked closely with Police Forces across the UK as consultants for animal mutilation cases, and we intend to forge closer links with the coastguard and other community services. We want to work closely with those who regularly travel into the Bristol Channel, so that if the recent trend of exotic animal visitors to our coastal waters continues, we can be out there as soon as possible.

Apart from having been the only Fortean Zoological organisation in the world to have consistently published material on all aspects of the subject for over a decade, we have achieved the following concrete results:

• Disproved the myth relating to the headless so-called sea-serpent carcass of Durgan beach in Cornwall 1975
• Disproved the story of the 1988 puma skull of

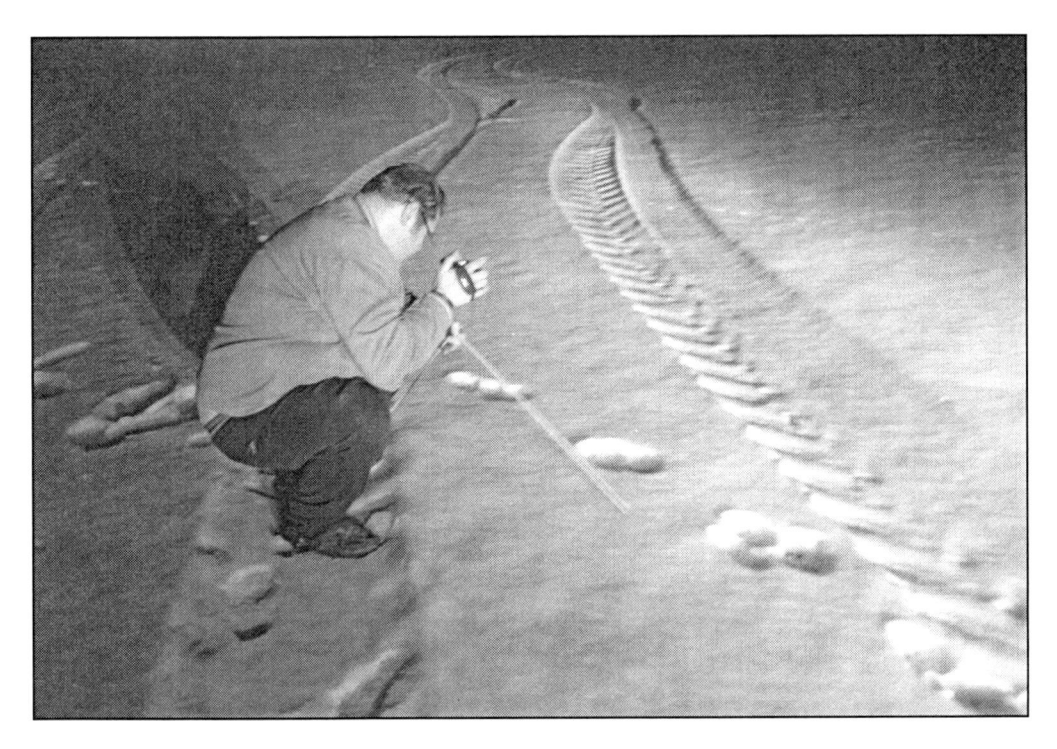

Lustleigh Cleave
- Carried out the only in-depth research ever into the mythos of the Cornish Owlman.
- Made the first records of a tropical species of lamprey
- Made the first records of a luminous cave gnat larva in Thailand
- Discovered a possible new species of British mammal - the beech marten
- In 1994-6 carried out the first archival fortean zoological survey of Hong Kong
- In the year 2000, CFZ theories were confirmed when a new species of lizard was added to the British List
- Identified the monster of Martin Mere in Lancashire as a giant wels catfish
- Expanded the known range of Armitage's skink in the Gambia by 80%
- Obtained photographic evidence of the remains of Europe's largest known pike
- Carried out the first ever in-depth study of the ninki-nanka
- Carried out the first attempt to breed Puerto Rican cave snails in captivity
- Were the first European explorers to visit the 'lost valley' in Sumatra
- Published the first ever evidence for a new tribe of pygmies in Guyana
- Published the first evidence for a new species of caiman in Guyana
- Filmed unknown creatures

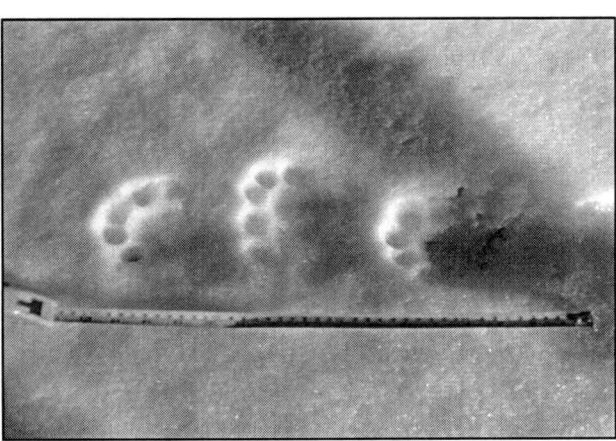

on a monster-haunted lake in Ireland for the first time
- Had a sighting of orang pendek in Sumatra in 2009
- Found leopard hair, subsequently identified by DNA analysis, from rural North Devon in 2010
- Brought back hairs which appear to be from an unknown primate in Sumatra
- Published some of the best evidence ever for the almasty in southern Russia

CFZ Expeditions and Investigations include:

- 1998 Puerto Rico, Florida, Mexico (Chupacabras)
- 1999 Nevada (Bigfoot)
- 2000 Thailand (Naga)
- 2002 Martin Mere (Giant catfish)
- 2002 Cleveland (Wallaby mutilation)

- 2003 Bolam Lake (BHM Reports)
- 2003 Sumatra (Orang Pendek)
- 2003 Texas (Bigfoot; giant snapping turtles)
- 2004 Sumatra (Orang Pendek; cigau, a sabre-toothed cat)
- 2004 Illinois (Black panthers; cicada swarm)
- 2004 Texas (Mystery blue dog)
- Loch Morar (Monster)
- 2004 Puerto Rico (Chupacabras; carnivorous cave snails)
- 2005 Belize (Affiliate expedition for hairy dwarfs)
- 2005 Loch Ness (Monster)
- 2005 Mongolia (Allghoi Khorkhoi aka Mongolian death worm)

- 2006 Gambia (Gambo - Gambian sea monster , Ninki Nanka and Armitage's skink
- 2006 Llangorse Lake (Giant pike, giant eels)
- 2006 Windermere (Giant eels)
- 2007 Coniston Water (Giant eels)
- 2007 Guyana (Giant anaconda, didi, water tiger)
- 2008 Russia (Almasty)
- 2009 Sumatra (Orang pendek)
- 2009 Republic of Ireland (Lake Monster)
- 2010 Texas (Blue Dogs)
- 2010 India (Mande Burung)
- 2011 Sumatra (Orang-pendek)

For details of current membership fees, current expeditions and investigations, and voluntary posts within the CFZ that need your help, please do not hesitate to contact us.

The Centre for Fortean Zoology,
Myrtle Cottage,
Woolfardisworthy,
Bideford, North Devon
EX39 5QR

Telephone 01237 431413
Fax +44 (0)7006-074-925
eMail info@cfz.org.uk

Websites:

www.cfz.org.uk
www.weirdweekend.org

THE WORLD'S WEIRDEST PUBLISHING COMPANY

HOW TO START A PUBLISHING EMPIRE

Unlike most mainstream publishers, we have a non-commercial remit, and our mission statement claims that "we publish books because they deserve to be published, not because we think that we can make money out of them". Our motto is the Latin Tag *Pro bona causa facimus* (we do it for good reason), a slogan taken from a children's book *The Case of the Silver Egg* by the late Desmond Skirrow.

WIKIPEDIA: "The first book published was in 1988. *Take this Brother may it Serve you Well* was a guide to Beatles bootlegs by Jonathan Downes. It sold quite well, but was hampered by very poor production values, being photocopied, and held together by a plastic clip binder. In 1988 A5 clip binders were hard to get hold of, so the publishers took A4 binders and cut them in half with a hacksaw. It now reaches surprisingly high prices second hand.

The production quality improved slightly over the years, and after 1999 all the books produced were ringbound with laminated colour covers. In 2004, however, they signed an agreement with Lightning Source, and all books are now produced perfect bound, with full colour covers."

Until 2010 all our books, the majority of which are/were on the subject of mystery animals and allied disciplines, were published by `CFZ Press`, the publishing arm of the Centre for Fortean Zoology (CFZ), and we urged our readers and followers to draw a discreet veil over the books that we published that were completely off topic to the CFZ.

However, in 2010 we decided that enough was enough and launched a second imprint, `Fortean Words` which aims to cover a wide range of non animal-related esoteric subjects. Other imprints will be launched as and when we feel like it, however the basic ethos of the company remains the same: Our job is to publish books and magazines that we feel are worth publishing, whether or not they are going to sell. Money is, after all - as my dear old Mama once told me - a rather vulgar subject, and she would be rolling in her grave if she thought that her eldest son was somehow in `trade`.

Luckily, so far our tastes have turned out not to be that rarified after all, and we have sold far more books than anyone ever thought that we would, so there is a moral in there somewhere...

Jon Downes,
Woolsery, North Devon
July 2010

Other Books in Print

Sea Serpent Carcasses—Scotland from the Stronsa Monster to Loch Ness by Glen Vaudrey
The CFZ Yearbook 2012 edited by Jonathan and Corinna Downes
ORANG PENDEK: Sumatra's Forgotten Ape by Richard Freeman
THE MYSTERY ANIMALS OF THE BRITISH ISLES: London by Neil Arnold
CFZ EXPEDITION REPORT: India 2010 by Richard Freeman *et al*
The Cryptid Creatures of Florida by Scott Marlow
Dead of Night by Lee Walker
The Mystery Animals of the British Isles: The Northern Isles by Glen Vaudrey
THE MYSTERY ANIMALS OF THE BRTISH ISLES: Gloucestershire and Worcestershire by Paul Williams
When Bigfoot Attacks by Michael Newton
Weird Waters – The Mystery Animals of Scandinavia: Lake and Sea Monsters by Lars Thomas
The Inhumanoids by Barton Nunnelly
Monstrum! A Wizard's Tale by Tony "Doc" Shiels
CFZ Yearbook 2011 edited by Jonathan Downes
Karl Shuker's Alien Zoo by Shuker, Dr Karl P.N
Tetrapod Zoology Book One by Naish, Dr Darren
The Mystery Animals of Ireland by Gary Cunningham and Ronan Coghlan
Monsters of Texas by Gerhard, Ken
The Great Yokai Encyclopaedia by Freeman, Richard
NEW HORIZONS: Animals & Men issues 16-20 Collected Editions Vol. 4 by Downes, Jonathan
A Daintree Diary -
Tales from Travels to the Daintree Rainforest in tropical north Queensland, Australia by Portman, Carl
Strangely Strange but Oddly Normal by Roberts, Andy
Centre for Fortean Zoology Yearbook 2010 by Downes, Jonathan
Predator Deathmatch by Molloy, Nick
Star Steeds and other Dreams by Shuker, Karl
CHINA: A Yellow Peril? by Muirhead, Richard
Mystery Animals of the British Isles: The Western Isles by Vaudrey, Glen

Giant Snakes - Unravelling the coils of mystery by Newton, Michael
Mystery Animals of the British Isles: Kent by Arnold, Neil
Centre for Fortean Zoology Yearbook 2009 by Downes, Jonathan
CFZ EXPEDITION REPORT: Russia 2008 by Richard Freeman *et al*, Shuker, Karl (fwd)
Dinosaurs and other Prehistoric Animals on Stamps - A Worldwide catalogue by Shuker, Karl P. N
Dr Shuker's Casebook by Shuker, Karl P.N
The Island of Paradise - chupacabra UFO crash retrievals, and accelerated evolution on the island of Puerto Rico by Downes, Jonathan
The Mystery Animals of the British Isles: Northumberland and Tyneside by Hallowell, Michael J
Centre for Fortean Zoology Yearbook 1997 by Downes, Jonathan (Ed)
Centre for Fortean Zoology Yearbook 2002 by Downes, Jonathan (Ed)
Centre for Fortean Zoology Yearbook 2000/1 by Downes, Jonathan (Ed)
Centre for Fortean Zoology Yearbook 1998 by Downes, Jonathan (Ed)
Centre for Fortean Zoology Yearbook 2003 by Downes, Jonathan (Ed)
In the wake of Bernard Heuvelmans by Woodley, Michael A
CFZ EXPEDITION REPORT: Guyana 2007 by Richard Freeman *et al*, Shuker, Karl (fwd)
Centre for Fortean Zoology Yearbook 1999 by Downes, Jonathan (Ed)
Big Cats in Britain Yearbook 2008 by Fraser, Mark (Ed)
Centre for Fortean Zoology Yearbook 1996 by Downes, Jonathan (Ed)
THE CALL OF THE WILD - Animals & Men issues 11-15
Collected Editions Vol. 3 by Downes, Jonathan (ed)
Ethna's Journal by Downes, C N
Centre for Fortean Zoology Yearbook 2008 by Downes, J (Ed)
DARK DORSET -Calendar Custome by Newland, Robert J
Extraordinary Animals Revisited by Shuker, Karl
MAN-MONKEY - In Search of the British Bigfoot by Redfern, Nick
Dark Dorset Tales of Mystery, Wonder and Terror by Newland, Robert J and Mark North
Big Cats Loose in Britain by Matthews, Marcus
MONSTER! - The A-Z of Zooform Phenomena by Arnold, Neil
The Centre for Fortean Zoology 2004 Yearbook by Downes, Jonathan (Ed)
The Centre for Fortean Zoology 2007 Yearbook by Downes, Jonathan (Ed)
CAT FLAPS! Northern Mystery Cats by Roberts, Andy
Big Cats in Britain Yearbook 2007 by Fraser, Mark (Ed)
BIG BIRD! - Modern sightings of Flying Monsters by Gerhard, Ken
THE NUMBER OF THE BEAST - Animals & Men issues 6-10
Collected Editions Vol. 1 by Downes, Jonathan (Ed)
IN THE BEGINNING - Animals & Men issues 1-5 Collected Editions Vol. 1 by Downes, Jonathan
STRENGTH THROUGH KOI - They saved Hitler's Koi and other stories by Downes, Jonathan
The Smaller Mystery Carnivores of the Westcountry by Downes, Jonathan
CFZ EXPEDITION REPORT: Gambia 2006 by Richard Freeman *et al*, Shuker, Karl (fwd)
The Owlman and Others by Jonathan Downes
The Blackdown Mystery by Downes, Jonathan

Big Cats in Britain Yearbook 2006 by Fraser, Mark (Ed)
Fragrant Harbours - Distant Rivers by Downes, John T
Only Fools and Goatsuckers by Downes, Jonathan
Monster of the Mere by Jonathan Downes
Dragons:More than a Myth by Freeman, Richard Alan
Granfer's Bible Stories by Downes, John Tweddell
Monster Hunter by Downes, Jonathan

Fortean Words

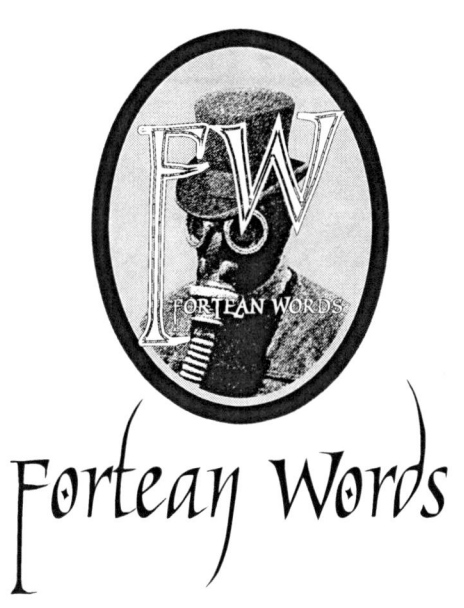

The Centre for Fortean Zoology has for several years led the field in Fortean publishing. CFZ Press is the only publishing company specialising in books on monsters and mystery animals. CFZ Press has published more books on this subject than any other company in history and has attracted such well known authors as Andy Roberts, Nick Redfern, Michael Newton, Dr Karl Shuker, Neil Arnold, Dr Darren Naish, Jon Downes, Ken Gerhard and Richard Freeman.

Now CFZ Press are launching a new imprint. Fortean Words is a new line of books dealing with Fortean subjects other than cryptozoology, which is - after all - the subject the CFZ are best known for. Fortean Words is being launched with a spectacular multi-volume series called *Haunted Skies* which covers British UFO sightings between 1940 and 2010. Former policeman John Hanson and his long-suffering partner Dawn Holloway have compiled a peerless library of sighting reports, many that have not been made public before.

Other books include a look at the Berwyn Mountains UFO case by renowned Fortean Andy Roberts and a series of forthcoming books by transatlantic researcher Nick Redfern. CFZ Press are dedicated to maintaining the fine quality of their works with Fortean Words. New authors tackling new subjects will always be encouraged, and we hope that our books will continue to be as ground-breaking and popular as ever.

Haunted Skies Volume One 1940-1959 by John Hanson and Dawn Holloway
Haunted Skies Volume Two 1960-1965 by John Hanson and Dawn Holloway
Haunted Skies Volume Three 1965-1967 by John Hanson and Dawn Holloway
Haunted Skies Volume Four 1968-1971 by John Hanson and Dawn Holloway
Haunted Skies Volume Five 1971-1974 by John Hanson and Dawn Holloway
Grave Concerns by Kai Roberts

Police and the Paranormal by Andy Owens
Dead of Night by Lee Walker
Space Girl Dead on Spaghetti Junction - an anthology by Nick Redfern
I Fort the Lore - an anthology by Paul Screeton
UFO Down - the Berwyn Mountains UFO Crash by Andy Roberts
The Grail by Ronan Coghlan

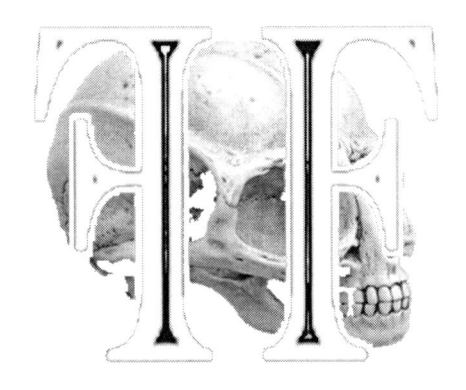

Fortean Fiction

Just before Christmas 2011, we launched our third imprint, this time dedicated to - let's see if you guessed it from the title - fictional books with a Fortean or cryptozoological theme. We have published a few fictional books in the past, but now think that because of our rising reputation as publishers of quality Forteana, that a dedicated fiction imprint was the order of the day.

We launched with four titles:

Green Unpleasant Land by Richard Freeman
Left Behind by Harriet Wadham
Dark Ness by Tabitca Cope
Snap! By Steven Bredice

Sea Serpent Carcasses Scotland

Artist's note

Readers may care to note that both original oils and colour prints of the cryptozoology art work of Glen Vaudrey are available for purchase. Contact the CFZ art team for details at Art@CFZ.org.uk